Getting Started with the SAS® 9.3 Output Delivery System

SAS® Documentation

The correct bibliographic citation for this manual is as follows: SAS Institute Inc. 2012. *Getting Started with the SAS® 9.3 Ouput Delivery System*. Cary, NC: SAS Institute Inc.

Getting Started with the SAS® 9.3 Ouput Delivery System

Copyright © 2012, SAS Institute Inc., Cary, NC, USA

All rights reserved. Produced in the United States of America.

For a hardcopy book: No part of this publication may be reproduced, stored in a retrieval system, or transmitted, in any form or by any means, electronic, mechanical, photocopying, or otherwise, without the prior written permission of the publisher, SAS Institute Inc.

For a Web download or e-book: Your use of this publication shall be governed by the terms established by the vendor at the time you acquire this publication.

The scanning, uploading, and distribution of this book via the Internet or any other means without the permission of the publisher is illegal and punishable by law. Please purchase only authorized electronic editions and do not participate in or encourage electronic piracy of copyrighted materials. Your support of others' rights is appreciated.

U.S. Government Restricted Rights Notice: Use, duplication, or disclosure of this software and related documentation by the U.S. government is subject to the Agreement with SAS Institute and the restrictions set forth in FAR 52.227–19, Commercial Computer Software-Restricted Rights (June 1987).

SAS Institute Inc., SAS Campus Drive, Cary, North Carolina 27513.

1st printing, August 2012

SAS® Publishing provides a complete selection of books and electronic products to help customers use SAS software to its fullest potential. For more information about our e-books, e-learning products, CDs, and hard-copy books, visit the SAS Publishing Web site at `support.sas.com/publishing` or call 1-800-727-3228.

SAS® and all other SAS Institute Inc. product or service names are registered trademarks or trademarks of SAS Institute Inc. in the USA and other countries. ® indicates USA registration.

Other brand and product names are registered trademarks or trademarks of their respective companies.

Contents

About This Book v
Accessibility for the SAS Output Delivery System ix
Recommended Reading xi

Chapter 1 • Introduction 1
    What Is the Output Delivery System? 1
    Benefits of Using ODS 2
    How Does ODS Work? 2
    Basic Usage 4
    ODS-Specific Windows 11
    Words to Know 14

Chapter 2 • Learning by Example: Creating Custom Reports with ODS 17
    About the Scenario in This Book 17
    Creating the Default Output 17

Chapter 3 • Creating an ODS Document 21
    About the Task That You Will Perform 21
    Creating an ODS Document 21
    For More Information 24

Chapter 4 • Selecting the Contents of Your Report 25
    About the Tasks That You Will Perform 25
    Identify the Output Objects 25
    Select the Output Objects 28
    For More Information 38

Chapter 5 • Integrating Output with Popular Business Applications and SAS ... 39
    About the Tasks That You Will Perform 39
    Creating RTF Output 40
    Creating PDF Output 42
    Creating Enhanced HTML Output 43
    Creating Excel Output 45
    Combined Program 45
    For More Information 48

Chapter 6 • Customizing the Presentation of a Report 49
    About the Tasks That You Will Perform 49
    Customized RTF Output 51
    Customized PROC TABULATE Output 52
    Customized PDF Output 54
    Customized HTML Output 56
    Customized Excel Output 57
    Combined Program 59
    For More Information 63

Chapter 7 • Next Steps: A Quick Look at Advanced Features 65
    Working with the TEMPLATE Procedure 65
    Working with ODS Graphics 70

Advanced Features of the DOCUMENT Procedure 74
ODS and the DATA Step 78
Where to Go from Here 80

***Index* 81**

About This Book

Audience

This book is intended for new or novice users of the SAS Output Delivery System. The documentation assumes familiarity with Base SAS programming and the SAS windowing environment. Although this familiarity is assumed, users who are not familiar will still be able to complete the tasks that are described in this book.

Syntax Conventions for the SAS Language

Overview of Syntax Conventions for the SAS Language

SAS uses standard conventions in the documentation of syntax for SAS language elements. These conventions enable you to easily identify the components of SAS syntax. The conventions can be divided into these parts:

- syntax components
- style conventions
- special characters
- references to SAS libraries and external files

Syntax Components

The components of the syntax for most language elements include a keyword and arguments. For some language elements, only a keyword is necessary. For other language elements, the keyword is followed by an equal sign (=).

keyword
  specifies the name of the SAS language element that you use when you write your program. Keyword is a literal that is usually the first word in the syntax. In a CALL routine, the first two words are keywords.

In the following examples of SAS syntax, the keywords are the first words in the syntax:

CHAR (*string, position*)

CALL RANBIN (*seed, n, p, x*);

ALTER (*alter-password*)

BEST *w*.

REMOVE <*data-set-name*>

In the following example, the first two words of the CALL routine are the keywords:

CALL RANBIN(*seed, n, p, x*)

The syntax of some SAS statements consists of a single keyword without arguments:

DO;
... *SAS code* ...
END;

Some system options require that one of two keyword values be specified:

DUPLEX | NODUPLEX

argument
: specifies a numeric or character constant, variable, or expression. Arguments follow the keyword or an equal sign after the keyword. The arguments are used by SAS to process the language element. Arguments can be required or optional. In the syntax, optional arguments are enclosed between angle brackets.

    In the following example, *string* and *position* follow the keyword CHAR. These arguments are required arguments for the CHAR function:

    CHAR (*string, position*)

    Each argument has a value. In the following example of SAS code, the argument *string* has a value of 'summer', and the argument *position* has a value of 4:`x=char('summer', 4);`

    In the following example, *string* and *substring* are required arguments, while *modifiers* and *startpos* are optional.

    FIND(*string, substring* <,*modifiers*> <,*startpos*>

Note: In most cases, example code in SAS documentation is written in lowercase with a monospace font. You can use uppercase, lowercase, or mixed case in the code that you write.

Style Conventions

The style conventions that are used in documenting SAS syntax include uppercase bold, uppercase, and italic:

UPPERCASE BOLD
: identifies SAS keywords such as the names of functions or statements. In the following example, the keyword ERROR is written in uppercase bold:

    ERROR<*message*>;

UPPERCASE
: identifies arguments that are literals.

    In the following example of the CMPMODEL= system option, the literals include BOTH, CATALOG, and XML:

    CMPMODEL = BOTH | CATALOG | XML

italics
: identifies arguments or values that you supply. Items in italics represent user-supplied values that are either one of the following:

- nonliteral arguments In the following example of the LINK statement, the argument *label* is a user-supplied value and is therefore written in italics:

  LINK *label*;

- nonliteral values that are assigned to an argument

  In the following example of the FORMAT statement, the argument DEFAULT is assigned the variable *default-format*:

  FORMAT = *variable-1* <, ..., *variable-nformat*><DEFAULT = *default-format*>;

Items in italics can also be the generic name for a list of arguments from which you can choose (for example, *attribute-list*). If more than one of an item in italics can be used, the items are expressed as *item-1, ..., item-n*.

Special Characters

The syntax of SAS language elements can contain the following special characters:

=
: an equal sign identifies a value for a literal in some language elements such as system options.

  In the following example of the MAPS system option, the equal sign sets the value of MAPS:

  MAPS = *location-of-maps*

< >
: angle brackets identify optional arguments. Any argument that is not enclosed in angle brackets is required.

  In the following example of the CAT function, at least one item is required:

  CAT (*item-1* <, ..., *item-n*>)

|
: a vertical bar indicates that you can choose one value from a group of values. Values that are separated by the vertical bar are mutually exclusive.

  In the following example of the CMPMODEL= system option, you can choose only one of the arguments:

  CMPMODEL = BOTH | CATALOG | XML

...
: an ellipsis indicates that the argument or group of arguments following the ellipsis can be repeated. If the ellipsis and the following argument are enclosed in angle brackets, then the argument is optional.

  In the following example of the CAT function, the ellipsis indicates that you can have multiple optional items:

  CAT (*item-1* <, ..., *item-n*>)

'*value*' or "*value*"
: indicates that an argument enclosed in single or double quotation marks must have a value that is also enclosed in single or double quotation marks.

  In the following example of the FOOTNOTE statement, the argument *text* is enclosed in quotation marks:

  FOOTNOTE <*n*> <*ods-format-options* '*text*' | "*text*">;

;
a semicolon indicates the end of a statement or CALL routine.

In the following example each statement ends with a semicolon: `data namegame; length color name $8; color = 'black'; name = 'jack'; game = trim(color) || name; run;`

References to SAS Libraries and External Files

Many SAS statements and other language elements refer to SAS libraries and external files. You can choose whether to make the reference through a logical name (a libref or fileref) or use the physical filename enclosed in quotation marks. If you use a logical name, you usually have a choice of using a SAS statement (LIBNAME or FILENAME) or the operating environment's control language to make the association. Several methods of referring to SAS libraries and external files are available, and some of these methods depend on your operating environment.

In the examples that use external files, SAS documentation uses the italicized phrase *file-specification*. In the examples that use SAS libraries, SAS documentation uses the italicized phrase *SAS-library*. Note that *SAS-library* is enclosed in quotation marks:

```
infile file-specification obs = 100;
libname libref 'SAS-library';
```

Accessibility for the SAS Output Delivery System

The Output Delivery System (ODS) conforms to Section 508 of the U.S. Rehabilitation Act guidelines for Web-based content. If you have specific questions about the accessibility of SAS products, send them to accessibility@sas.com or call SAS Technical Support.

The following additional accessibility items are available as programming options:

Event Variables

> **TIP** For information about the following event variables, see the "Event Variables" in Chapter 15 of *SAS Output Delivery System: User's Guide*.

ABBR
: specifies an abbreviation for an event variable.

ACRONYM
: specifies an acronym for an event variable.

ALT
: specifies an alternate description of an event variable.

CAPTION
: specifies a caption for a table.

LONGDESC
: specifies a long description of an event variable.

SUMMARY
: specifies a summary of a table.

Style Template

STYLES.HIGHCONTRAST
: creates the same output as the default output except all of the colors are black on white.

Header Attributes

> **TIP** For information about the following header attributes, see the "Event Variables" in Chapter 15 of *SAS Output Delivery System: User's Guide*.

ABBR=
: specifies an abbreviation for a header.

ACRONYM=
: specifies an acronym for a header.

ALT=
: specifies an alternate description of a header.

GENERIC
: specifies whether multiple columns can use the same header.

LONGDESC=
: specifies a long description of a header.

Table Attibutes

LONGDESC=
: specifies a long description of a table.

ALT=
: specifies an alternate description of a table.

The following tagsets and ODS statements create output that is 508 compliant:

ODS PHTML Statement
: opens, manages, or closes the PHTML destination, which produces simple HTML output that uses 12 style elements and no class attributes. For more information about the ODS PHTML statement, see the "ODS PHTML Statement" in *SAS Output Delivery System: User's Guide*.

ODS HTMLCSS Statement
: opens, manages, or closes the HTMLCSS destination, which produces HTML output with cascading style sheets (CSS). For more information about the ODS HTMLCSS statement, see the "ODS HTMLCSS Statement" in *SAS Output Delivery System: User's Guide*.

ODS HTML Statement
: opens, manages, or closes the HTML destination, which produces HTML 4.0 output that contains embedded style sheets. For more information about the ODS HTML statement, see the "ODS HTML Statement " in *SAS Output Delivery System: User's Guide*.

MSOFFICE2K Tagset
: produces HTML code for output generated by ODS for Microsoft Office products. For more information about the MSOFFICE2K tagset, see "MSOFFICE2K" in Chapter 6 of *SAS Output Delivery System: User's Guide*.

In SAS 9.1 and later releases, all of the accessibility enhancements have been merged into the ODS HTML tagsets. No additional steps are required.

Recommended Reading

Here is the recommended reading list for this title. For a complete list of SAS publications, go to `http://support.sas.com/publishing/index.html`.

- *SAS Output Delivery System: User's Guide*
- *Base SAS Procedures Guide*
- *SAS Language Reference: Concepts*
- *SAS Data Set Options: Reference*
- *SAS Functions and CALL Routines: Reference*
- *SAS Statements: Reference*
- *SAS System Options: Reference*
- *Step-by-Step Programming with Base SAS Software*

The recommended reading list from **SAS Press** includes:

- The Little SAS Book: A Primer, Revised Second Edition
- *Output Delivery System: The Basics and Beyond*
- *Output Delivery System: The Basics*
- *Instant ODS: Style Templates for the SAS Output Delivery System*

For a complete list of SAS publications, go to support.sas.com/bookstore. If you have questions about which titles you need, please contact a SAS Publishing Sales Representative:

SAS Publishing Sales
SAS Campus Drive
Cary, NC 27513-2414
Phone: 1-800-727-3228
Fax: 1-919-677-8166
E-mail: sasbook@sas.com
Web address: support.sas.com/bookstore

Chapter 1
Introduction

What Is the Output Delivery System? 1
Benefits of Using ODS 2
How Does ODS Work? 2
    Components of ODS 2
    Where Does ODS Put My Output? 4
Basic Usage 4
    Overview 4
    The DOCUMENT Procedure 4
    ODS Global Statements 5
    Base SAS Reporting Procedures 9
    TEMPLATE Procedure 10
ODS-Specific Windows 11
    Overview 11
    The Documents Window 11
    The Templates Window 12
    The Template Browser Window 13
Words to Know 14

What Is the Output Delivery System?

The Output Delivery System (ODS) enables you to customize the content of your output, choose how your output is formatted, and customize the appearance of your output. Examples of output formats are HTML, PDF, RTF, and LISTING.

Important features of ODS include the following:

- ODS destinations produce the following types of output:
  - traditional monospace output
  - an output data set
  - an ODS document that contains a hierarchical file of the output objects
  - a zip file
  - output such as PostScript and PDF that is formatted for a high-resolution printer
  - output that is formatted in a markup language such as HTML and XML
  - RTF output that is formatted for use with Microsoft Word

- ODS provides table templates that define the structure of the output from SAS procedures and from the DATA step. You can customize the output by modifying these table templates or by creating your own.

- ODS enables you to choose an individual output object to be formatted in a different way. For example, PROC UNIVARIATE produces multiple output objects. You can easily create HTML output, one or more output data sets, LISTING output, or printer output from one or all of these output objects.

- ODS stores a link to each output object in the Results window of the SAS windowing environment.

- SAS output formatting is now centralized in ODS. When ODS destinations are added, they are automatically made available to the DATA step and all procedures that support ODS.

- ODS enables you to produce output for numerous business applications from a single source. This feature saves you time and system resources because you can produce multiple types of output with a single run of your procedure or data query. For example, you can create output formatted for Microsoft Excel, Adobe Acrobat, and Microsoft Word.

Benefits of Using ODS

ODS gives you the flexibility to generate, format, and reproduce SAS procedure and DATA step output. You can use ODS to accomplish the following tasks:

Create reports for popular software applications.
: With ODS, you can use ODS destination statements to create output specifically for software other than SAS and make that output easy to access. For example, you can use the ODS PDF statement to create PDF files for viewing with Adobe Acrobat or for printing. You can use the ODS HTML statement to create output for the Web. The ODS RTF statement creates output for Microsoft Word. For complete documentation on the ODS destination statements, see Chapter 6, "Dictionary of ODS Language Statements," in *SAS Output Delivery System: User's Guide*.

Customize the report contents.
: ODS enables you to modify the contents of your output. With ODS, you can embed graphics, select specific cell contents to display, and create embedded links in tables and graphs. You can select specific tables or graphs from procedure output to print or you can exclude them. You can create SAS data sets directly from tables or graphics.

Customize the presentation.
: ODS enables you to change the appearance of your output. You can change the colors, fonts, and borders of your output. You can customize the layout, format, headers, and style. You can add images and embedded URLs.

How Does ODS Work?

Components of ODS

ODS creates various types of tabular output by combining raw data with one or more table templates to produce one or more output objects. The basic component of ODS functionality is the output object. The PROC or DATA step that you run provides the data component (raw data) and the name of the table template that contains the formatting instructions. The data component and table template together form the output object. There are two types of output objects created by ODS: tabular output objects and graphical output objects. These objects can be sent to any or all ODS destinations, such as PDF, HTML, RTF, or LISTING. By default, in the SAS windowing environment for Windows and UNIX, SAS uses ODS to produce HTML output. By default, in batch mode, SAS produces LISTING output. By specifying an ODS destination, you control the type of output that SAS creates.

You can use ODS to manipulate one or more output objects in many different ways.

- You can use the DOCUMENT procedure to select, rearrange, store, or replay output objects.
- You can use ODS output destinations to create output in many different formats.
- You can use the ODS TRACE statement to determine what output objects are available, and you can use the ODS SELECT or ODS EXCLUDE statement to select or exclude the output object from an output destination.

The following figure shows how SAS produces ODS output.

Figure 1.1 ODS Processing: What Goes In and What Comes Out

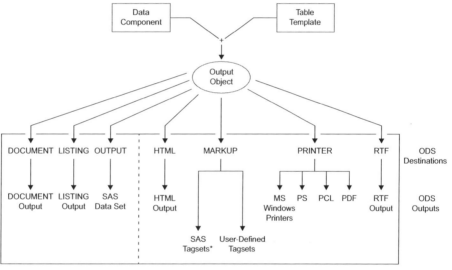

Table 1.1 *List of Tagsets That SAS Provides and Supports

CHTML	CSV	CSVALL	CSVBYLINE
DEFAULT	DOCBOOK	EXCELXP	HTML4
HTMLCSS	HTMLPANEL	IMODE	MSOFFICE2K
PHTML	PYX	RTF	SASREPORT
WML	WMLOLIST	XHTML	

Table 1.2 *Diagnostic Tagsets That SAS Supports*

EVENT_MAP	NAMEDHTML	SHORT_MAP	STYLE_DISPLAY
STYLE_POPUP	TEXT_MAP	TPL_STYLE_LIST	TPL_SYLE_MAP

Note: There are also preproduction tagsets. These tagsets can be found at `http://support.sas.com`. They are not yet supported by SAS Technical Support.

Where Does ODS Put My Output?

By default, SAS stores output created by ODS in your Work directory. In the SAS windowing environment for Windows and UNIX, after you have opened and closed the HTML destination, your output goes to your current working directory. You can use the ODS PREFERENCE statement anytime during your SAS session to return to the default behavior. This action is helpful when you are creating multiple graphics and do not want them to accumulate in your current working directory.

Basic Usage

Overview

ODS is used by all SAS software. However, you can explicitly use ODS with the following:

- DOCUMENT Procedure
- ODS Global Statements
- Base SAS Reporting Procedures
- TEMPLATE Procedure

The DOCUMENT Procedure

The combination of the ODS DOCUMENT statement and the DOCUMENT procedure enables you to store a report's individual components, and then modify and replay the report. The ODS DOCUMENT statement stores the actual ODS objects that are created when you run a report. You can then use PROC DOCUMENT to rearrange, duplicate, or remove output from the results of a procedure or a data query without invoking the procedures from the original report. You can also use PROC DOCUMENT to do the following:

- transform a report without rerunning an analysis or repeating a data query
- modify the structure of output
- display output to any ODS output format without executing the original procedure or DATA step
- navigate the current directory and list entries
- open and list ODS documents

- manage output
- store the ODS output objects in raw form

    Note: The output is kept in the original internal representation as a data component plus a table template.

The DOCUMENT destination has a graphical user interface (GUI), called the Documents window, for performing tasks. However, you can perform the same tasks with batch statement syntax using the DOCUMENT procedure.

For complete documentation on the DOCUMENT procedure, see Chapter 8, "The DOCUMENT Procedure," in *SAS Output Delivery System: User's Guide*.

ODS Global Statements

ODS global statements provide greater flexibility to generate, customize, and reproduce SAS procedure and DATA step output. You can use ODS global statements to control different features of ODS. ODS statements can be used anywhere in your SAS program. Some ODS statements remain in effect until you explicitly change them. Other ODS statements are automatically cleared. For complete documentation on ODS global statements, see the chapter about ODS statements in the *SAS Output Delivery System: User's Guide*.

ODS global statements are organized into two types.

Output Control Statements
: are statements that provide descriptive information about the specified output objects, and they indicate whether the style definition or table template is provided by SAS. The ODS EXCLUDE, ODS SELECT, and ODS TRACE statements are examples of output control statements.

    Output control statements can do the following:

    - select specific output objects for specific destinations
    - exclude specific output objects from specific destinations
    - specify the location where you want to search for or store style definitions or table templates
    - verify whether you are using a style definition or a table template that is provided by SAS
    - provide descriptive information about each specified output object, such as its name, label, template, path, and label path

ODS Destination (Report) Statements
: are statements that enable you to create output that is formatted for third-party software, such as HTML, RTF, and PDF. Or, they enable you to create output that is specific to SAS, such as an ODS document, LISTING output, or a SAS data set.

You can use ODS destination statements to generate and modify reports in formats such as HTML, XML, PDF, PostScript, RTF, and Microsoft Excel. The form for an ODS destination statement is the ODS statement block, which consists of ODS statements that open and close one or more ODS destinations sandwiched around your program. Your results are sent to one or more output destinations.

You can use one or more ODS destination statements, one or more PROC or DATA steps, and an ODS CLOSE statement to form an ODS statement block. An ODS block has the following form:

ODS *output-destination 1 <options(s)>*;

...

ODS *output-destination (n) <options(s)>*

<your SAS program>

ODS *destination close statement 1*;

...

ODS *destination close statement (n)*

In the ODS block, *output-destination* is the name of a valid ODS destination and *option(s)* are options that are valid for that destination. Your SAS program is inserted between the beginning ODS destination statement and the ODS CLOSE statement.

In the following example, the output from PROC PRINT and PROC CONTENTS is sent to the PDF and RTF destinations. The STYLE= option specifies what table template to apply to the output. By default, the PDF opens in Adobe Acrobat and the RTF opens in Microsoft Word.

```
options obs=10 nodate;
ods pdf file="myPdf.pdf" style=Banker;
ods rtf file="myRTF.rtf" style=BarrettsBlue text="RTF Output";
proc print data=sashelp.class;
run;

proc contents data=sashelp.class;
run;

ods pdf close;
ods rtf close;
```

Output 1.1   Default PDF Output

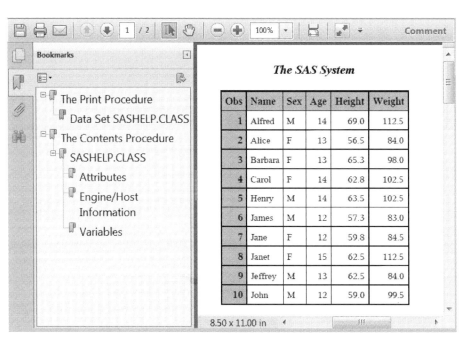

Output 1.2 PDF Output with Banker Style Applied

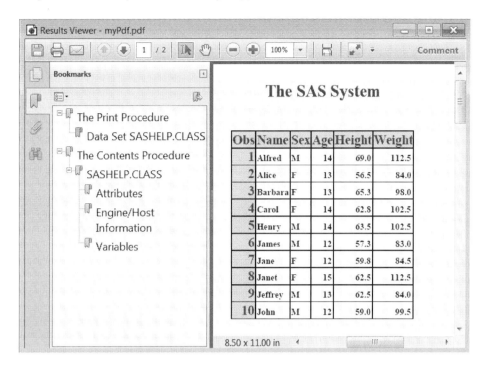

Output 1.3 Default RTF Output

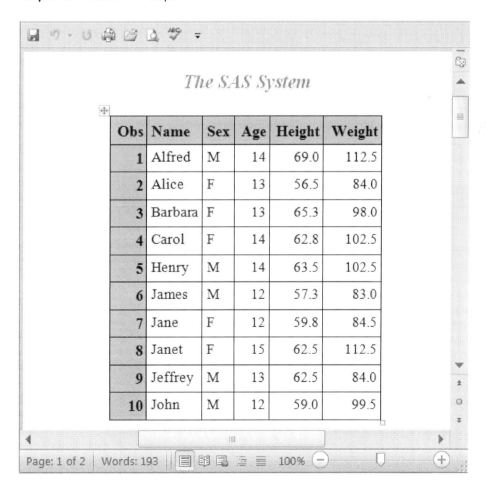

Output 1.4 RTF Output with BarrettsBlue Style Applied

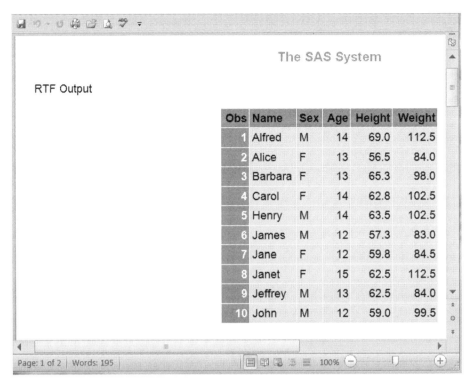

ODS destinations are organized into two categories.

Destinations Formatted by SAS
  These destinations produce output that is controlled and interpreted by SAS, such as a SAS data set, LISTING output, or an ODS document.

Destinations Formatted by a Third Party
  These destinations produce output that enables you to apply styles or markup languages. You can print to physical printers using page description languages. For example, you can produce output in PostScript, HTML, XML, or in a markup language that you created.

The following table lists the ODS destination categories, the destinations that each category includes, and the formatted output that results from each destination.

Table 1.3 Destination Category Table

Category	Destinations	Results
Formatted by SAS	DOCUMENT	ODS document
	LISTING	SAS LISTING output
	OUTPUT	SAS data set
Formatted by a Third Party	HTML	HTML file for online viewing
	MARKUP	Markup language tagsets

Category	Destinations	Results
	PRINTER	Printable output in one of three different formats: PCL, PDF, or PS (PostScript)
	RTF	Output written in Rich Text Format for use with Microsoft Word 2000

As destinations are added to ODS, they will automatically become available to the DATA step and all procedures that support ODS.

Base SAS Reporting Procedures

The Base SAS reporting procedures, PROC PRINT, PROC REPORT, and PROC TABULATE, enable you to quickly analyze your data and organize it into easy-to-read tables. You can use ODS options with the reporting procedures to give your report another dimension of expression and usability. For example, you can use the STYLE option with a PROC PRINT, PROC REPORT, or PROC TABULATE statement to change the appearance of your report. The following program uses the ODS STYLE option to create the colors in the output below:

```
Title "Height and Weight by Gender and Age";
proc report nowd data=sashelp.class
   style(header)=[background=white];
   col age (('gender' sex),(weight height));
   define age / style(header)=[background=lightgreen];
   define sex / across style(header)=[background=yellow] ' ';
   define weight / style(header)=[background=orange];
   define height / style(header)=[background=tan];
run;
```

Output 1.5 PROC REPORT Output with Styles Applied

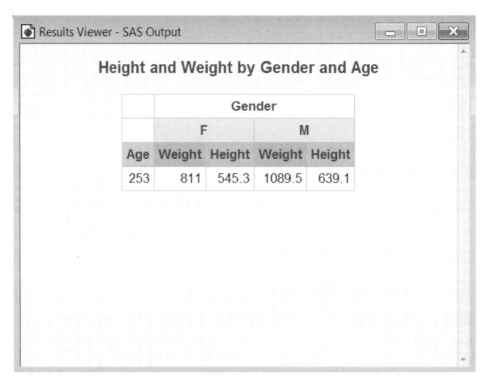

For complete documentation on the styles and style attributes that you can use with PROC PRINT, PROC TABULATE, and PROC REPORT, see the *Base SAS Procedures Guide*.

TEMPLATE Procedure

All SAS procedures produce output objects that ODS delivers to various ODS destinations based on the default specifications for the procedure or based on your own specifications. Output objects are commonly displayed as tables, data sets, or graphs. Each output object has an associated template provided by SAS that defines its presentation format. You can use the TEMPLATE procedure to view or alter a template or to create a new template by changing the headers, formats, column order, and so on.

The TEMPLATE procedure enables you to create or modify a template that you can apply to your output. You can also use the TEMPLATE procedure to navigate and manage the templates stored in template stores. ODS then uses these templates to produce formatted output. Using the TEMPLATE procedure is an advanced technique. For more information about advanced ODS techniques, see Chapter 7, "Next Steps: A Quick Look at Advanced Features ," on page 65.

For complete documentation on the TEMPLATE procedure, see the TEMPLATE procedure in the *SAS Output Delivery System: User's Guide*.

ODS-Specific Windows

Overview

There are three ODS windows that enable you to manipulate or browse your ODS output and templates.

- Documents Window
- Template Window
- Template Browser Window

The Documents Window

The Documents window displays ODS documents in a hierarchical tree structure.

The Documents window does the following:

- displays all ODS documents, including ODS documents stored in SAS libraries
- organizes, manages, and customizes the layout of the entries contained in ODS documents
- displays the property information of ODS documents
- replays entries
- renames, copies, moves, or deletes ODS documents
- creates shortcuts to ODS documents

To open the Documents window, do one of the following:

- Select **Results** in the Results window, and then select **View** ⇨ **Documents** from the taskbar.
- Right-click **Results** in the Results window, and then select **Documents**.
- Issue the following command on the command line in the SAS windowing environment:

    odsdocuments

This display shows a Documents window that contains an ODS document named Sasuser.Univ. In the display, notice that Sasuser.Univ contains several directory levels. The Exponential_x directory contains the Exp output object. When you double-click on an output object such as Exp, that output object is replayed in the Results window and sent to all open destinations.

Display 1.1 Documents Window

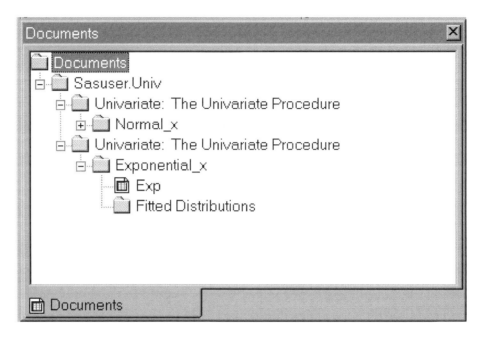

The Templates Window

Templates contain descriptive information that helps ODS determine the layout of your output. The Templates window enables you to manage all of the templates that are currently available to SAS. Specifically, you can use the Templates window to perform the following tasks:

- Browse ODS styles.
- View template properties.
- Browse template stores and item stores.
- Browse PROC TEMPLATE source code.

To open the Templates window, do one of the following:

- Select **Results** in the Results window, and then select **View** ⇨ **Templates** from the taskbar.
- Right-click **Results** in the Results window, and then select **Templates**.
- Issue the following command on the command line in the SAS windowing environment:

    odstemplates

The hierarchal view on the left side of the Templates window lists the item stores, template stores, directories, and items. The contents of a selected store or directory are displayed on the right side of the window.

Figure 1.2 Templates Window Showing Item Stores, Template Stores, Directories, and Items

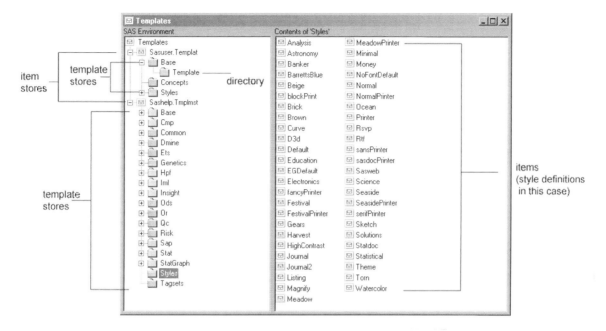

A template store is an item store that contains items that were created by the TEMPLATE procedure. Items that SAS provides are in the Sashelp.Tmplmst item store. By default, compiled templates are stored physically in the Sasuser.Templat item store. You can store items that you create in any template store where you have Write access. A template store can contain multiple levels (directories).

The Template Browser Window

The Template Browser window enables you to view the source code of a template. To open the Template Browser window, double-click on a template or style.

In the following display, the source code for the HTMLBlue style is shown in the Template Browser window.

Display 1.2 Templates Window and Template Browser Window

Words to Know

data components
: a form, similar to a SAS data set, that contains the results (numbers and characters) of a DATA step or PROC step.

item store
: a SAS library member that consists of pieces of information that can be accessed independently. The contents of an item store are organized in a directory tree structure, which is similar to the directory structure that is used by UNIX System Services or Windows. For example, a value might be stored in and located by using a directory path (`root_dir/sub_dir/value`). The SAS registry, Sasuser.Templat, and Sashelp.Tmplmst are examples of item stores.

template store
: an item store that contains items that were created by the TEMPLATE procedure. Items that SAS provides are in the Sashelp.Tmplmst item store.

ODS destination
: a designation that ODS uses to generate a specific type of output. For example, HTML, XML, LISTING, PostScript, RTF, and SAS data set are ODS destinations.

ODS document
: a hierarchy of output objects created by the DOCUMENT procedure. These output objects are unformatted and placed in a SAS item store.

ODS template
: a definition of how output should appear when it is formatted. An ODS template is stored as a compiled entry in a template store, which is also known as an item store. Common ODS template types include STATGRAPH, STYLE, CROSSTABS, TAGSET, and TABLE.

output object
: data component that is generated by a PROC or DATA step. It can also contain a table template that provides formatting instructions for the data.

table template
: a template that describes how to format the output for a tabular output object. A table template determines the order of table headers and footers, the order of columns, and the overall appearance of the output object. Each table template contains or references table elements.

Chapter 2
Learning by Example: Creating Custom Reports with ODS

About the Scenario in This Book 17
Creating the Default Output 17

About the Scenario in This Book

This book presents an ODS example that is intended to familiarize you with some of the basic features of ODS. From this example, you will learn tasks that will help you enhance and customize SAS output. After this chapter, you do not have to follow the chapters and steps in order.

For the purpose of the scenario in this book, you are a data analyst at a global furniture company. Your manager has asked you for various reports on the data, which you have created. However, you would like to present the information in a more visually pleasing way, and you want to display only the information that is required. This is the first of several presentations of the data, so you would like to save the information in a way that is easy to get to and modify if changes are needed. Your SAS session is being run in the SAS windowing environment in Windows.

Creating the Default Output

The following program creates the original PROC TABULATE, PROC UNIVARIATE, and PROC SGPANEL output. When you run this example program, you are creating ODS output. By default, HTML is created when you run code in the SAS windowing environment for Windows or UNIX. Your output (including your graphics) is sent to your current working directory. This output is viewable in the Results Viewer.

```
options nodate nonumber;
proc sort data=sashelp.prdsale out=prdsale;
    by Country;
run;

proc tabulate data=prdsale;
class region division prodtype;
classlev region division prodtype;
var actual;
```

```
       keyword all sum;
       keylabel all='Total';
       table (region all)*(division all),
             (prodtype all)*(actual*f=dollar10.) /
             misstext=[label='Missing']
             box=[label='Region by Division and Type'];

       title 'Actual Product Sales';
       title2 '(millions of dollars)';
       run;
       proc univariate data=prdsale;
       by Country;
       var actual;
       run;
       title 'Sales Figures for First Quarter by Product';
       proc sgpanel data=prdsale;
           where quarter=1;
           panelby product / novarname;
           vbar region / response=predict;
           vline region / response=actual lineattrs=GraphFit;
           colaxis fitpolicy=thin;
           rowaxis label='Sales';
       run;
```

When you create this output, you determine that there is more information than you want. Specifically, you need only the Extreme Observations, Quantiles, and Moments tables from the PROC UNIVARIATE output. In addition, you want to make the output easier to read. The steps in the following chapters show you how to accomplish these tasks.

Output 2.1 Default Output

Actual Product Sales
(millions of dollars)

Region by Division and Type		Product type		Total
		FURNITURE	OFFICE	
		Actual Sales	Actual Sales	Actual Sales
		Sum	Sum	Sum
Region	Division			
EAST	CONSUMER	$72,570	$108,686	$181,256
	EDUCATION	$73,901	$115,104	$189,005
	Total	$146,471	$223,790	$370,261
WEST	Division			
	CONSUMER	$76,209	$105,020	$181,229
	EDUCATION	$67,945	$110,902	$178,847
	Total	$144,154	$215,922	$360,076
Total	Division			
	CONSUMER	$148,779	$213,706	$362,485
	EDUCATION	$141,846	$226,006	$367,852
	Total	$290,625	$439,712	$730,337

Actual Product Sales
(millions of dollars)

The UNIVARIATE Procedure

Chapter 3
Creating an ODS Document

About the Task That You Will Perform 21
Creating an ODS Document 21
For More Information 24

About the Task That You Will Perform

You want to be able to modify your default output later without rerunning your procedures. To accomplish this, you can create an ODS document that contains output from the procedures. The combination of the ODS DOCUMENT statement and the DOCUMENT procedure enables you to store a report's individual components, and then modify and replay the report at a later time. The ODS DOCUMENT statement stores the actual ODS output objects that are created when running a report. You can then use the DOCUMENT procedure to rearrange, duplicate, or remove output from the results without rerunning the procedures from the original report.

Creating an ODS Document

Creating the ODS document with the ODS DOCUMENT statement is the first step toward ODS document functionality. The ODS DOCUMENT statement has the following form:

ODS DOCUMENT *action* | *option*;

In the ODS DOCUMENT statement below, the NAME= option assigns the document the name PrdDocument, and the WRITE option specifies that the document has Write access. Note that using Write access will overwrite existing documents in the library. You must always specify an ODS *destination* CLOSE statement when using ODS global statements. The ODS DOCUMENT CLOSE statement below closes the document so that it can be viewed in the Documents window.

The Documents window displays ODS documents in a hierarchical tree structure. To open the Documents window, issue the following command on the command line:
`odsdocuments`

```
    proc sort data=sashelp.prdsale out=prdsale;
        by Country;
    run;
```

```
ods document name=work.prddocument(write);

proc tabulate data=prdsale;
class region division prodtype;
classlev region division prodtype;
var actual;
keyword all sum;
keylabel all='Total';
table (region all)*(division all),
        (prodtype all)*(actual*f=dollar10.) /
        misstext=[label='Missing']
        box=[label='Region by Division and Type'];

title 'Actual Product Sales';
title2 '(millions of dollars)';
run;

proc univariate data=prdsale;
by Country;
var actual;
run;
title 'Sales Figures for First Quarter by Product';
proc sgpanel data=prdsale;
    where quarter=1;
    panelby product / novarname;
    vbar region / response=predict;
    vline region / response=actual lineattrs=GraphFit;
    colaxis fitpolicy=thin;
    rowaxis label='Sales';
run;
ods document close;
```

Display 3.1 Documents Window

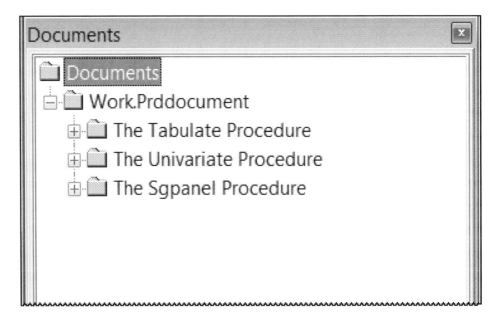

Display 3.2 Documents Window Expanded

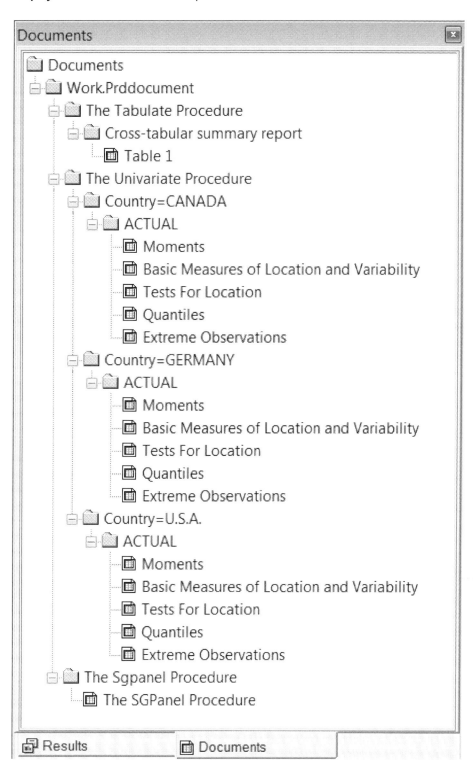

For More Information

For more information, see:

- "ODS DOCUMENT Statement" in *SAS Output Delivery System: User's Guide*.
- Chapter 8, "The DOCUMENT Procedure," in *SAS Output Delivery System: User's Guide*.

Chapter 4
Selecting the Contents of Your Report

About the Tasks That You Will Perform 25
Identify the Output Objects 25
    Using the ODS TRACE Statement 25
    Using the Documents Window 27

Select the Output Objects 28
    Using ODS Statements 28
    Using the Documents Window 29
    Comparing the Two Methods 38

For More Information 38

About the Tasks That You Will Perform

Your content is parceled out into output objects. For your presentation, you do not need all of the output. You want to display the following:

- All of the PROC TABULATE output.
- The Extreme Observations table, Quantiles table, and Moments table for Canada, Germany, and the United States from PROC UNIVARIATE.
- All of the PROC SGPANEL output.

There are two steps to select or exclude an output object. First, you must identify the name, label, or path of the output object. Second, you must use ODS to select or exclude the output object.

Identify the Output Objects

Using the ODS TRACE Statement

The easiest way to identify all of your output objects is with the ODS TRACE statement. Because the ODS TRACE statement is a global statement, you can place it anywhere in your program. By using the ODS TRACE statement, you can see all of the output objects at a glance. Because you want to select or exclude output objects created only by PROC UNIVARIATE, place the ODS TRACE ON statement before the PROC

UNIVARIATE step. Place the ODS TRACE OFF statement after the PROC UNIVARIATE step to stop the generation of trace information.

```
ods trace on;
proc univariate data=prdsale;
    by Country;
    var actual;
    run;
ods trace off;
```

Display 4.1 *Trace Output Viewed in the SAS Log*

```
Log - (Untitled)
56      ods trace on;
57      proc univariate data=prdsale;
58          by Country;
59          var actual;
60          run;

Output Added:
-------------
Name:       Moments
Label:      Moments
Template:   base.univariate.Moments
Path:       Univariate.ByGroup1.ACTUAL.Moments
-------------

Output Added:
-------------
Name:       BasicMeasures
Label:      Basic Measures of Location and Variability
Template:   base.univariate.Measures
Path:       Univariate.ByGroup1.ACTUAL.BasicMeasures
-------------

Output Added:
-------------
Name:       TestsForLocation
Label:      Tests For Location
Template:   base.univariate.Location
Path:       Univariate.ByGroup1.ACTUAL.TestsForLocatio
-------------

Output Added:
-------------
Name:       Quantiles
Label:      Quantiles
Template:   base.univariate.Quantiles
Path:       Univariate.ByGroup1.ACTUAL.Quantiles
-------------

Output Added:
-------------
Name:       ExtremeObs
Label:      Extreme Observations
Template:   base.univariate.ExtObs
Path:       Univariate.ByGroup1.ACTUAL.ExtremeObs
-------------
NOTE: The above message was for the following BY group
      Country=CANADA
```

Using the Documents Window

If you have created an ODS document that contains all of the procedure output, you can view the labels of the output objects in the Documents window.

Display 4.2 Output Objects Viewed in the Documents Window

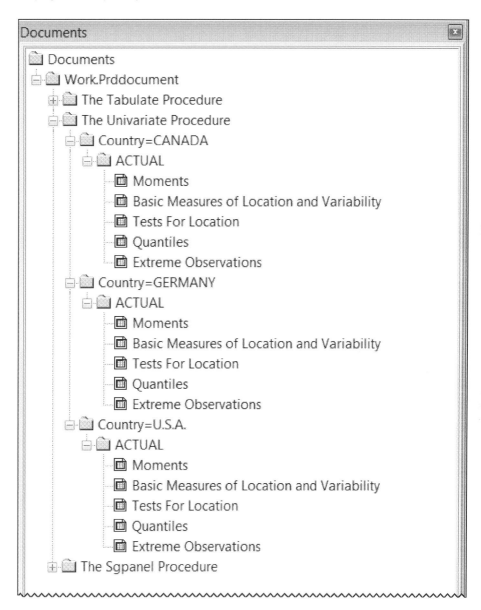

In the figure below, you can see that the labels of the output objects in the trace output correspond to the names given to the output objects in the Documents window.

Display 4.3 Comparing the Documents Window and the Trace Output

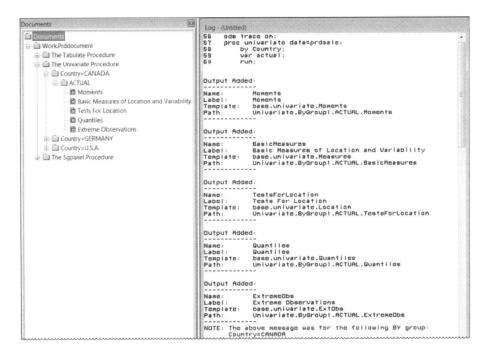

Select the Output Objects

Using ODS Statements

Once you have identified all of your output objects, you can use the ODS SELECT statement or the ODS EXCLUDE statement to select or exclude output objects. You can use the name, label, or path to specify the output object in either of these ODS statements.

```
proc sort data=sashelp.prdsale out=prdsale;
   by Country;
run;
ods html file='your-file-path\HTMLPrdhtml.html';
ods document name=work.prddocument(write);
proc tabulate data=prdsale;
class region division prodtype;
classlev region division prodtype;
var actual;
keyword all sum;
keylabel all='Total';
table (region all)*(division all),
      (prodtype all)*(actual*f=dollar10.) /
      misstext=[label='Missing']
      box=[label='Region by Division and Type'];

title 'Actual Product Sales';
```

```
    title2 '(millions of dollars)';
run;
ods select ExtremeObs Quantiles Moments;

proc univariate data=prdsale;
    by Country;
    var actual;
run;
proc sgpanel data=prdsale;
    where quarter=1;
    panelby product / novarname;
    vbar region / response=predict;
    vline region / response=actual lineattrs=GraphFit;
    colaxis fitpolicy=thin;
    rowaxis label='Sales';
run;
ods html close;
ods document close;
```

You can use the ODS EXCLUDE statement instead of the ODS SELECT statement. The following ODS EXCLUDE statement gives you the same results:

```
ods exclude BasicMeasures TestsForLocation;
```

Using the Documents Window

If you want to use the Documents window to select your output objects and to store your updated output, you must create a new ODS document.

In a previous chapter, you created an original ODS document. You can recall that ODS document.

```
proc sort data=sashelp.prdsale out=prdsale;
    by Country;
run;

ods document name=work.prddocument(write);

proc tabulate data=prdsale;
class region division prodtype;
classlev region division prodtype;
var actual;
keyword all sum;
keylabel all='Total';
table (region all)*(division all),
      (prodtype all)*(actual*f=dollar10.) /
      misstext=[label='Missing']
      box=[label='Region by Division and Type'];

title 'Actual Product Sales';
title2 '(millions of dollars)';
run;

proc univariate data=prdsale;
by Country;
var actual;
run;
```

```
title 'Sales Figures for First Quarter by Product';
proc sgpanel data=prdsale;
    where quarter=1;
    panelby product / novarname;
    vbar region / response=predict;
    vline region / response=actual lineattrs=GraphFit;
    colaxis fitpolicy=thin;
    rowaxis label='Sales';
run;
ods document close;
```

To create a new ODS document, right-click the **Documents** folder at the top of the Documents window, and then select **New Document**.

Display 4.4 Creating a New ODS Document

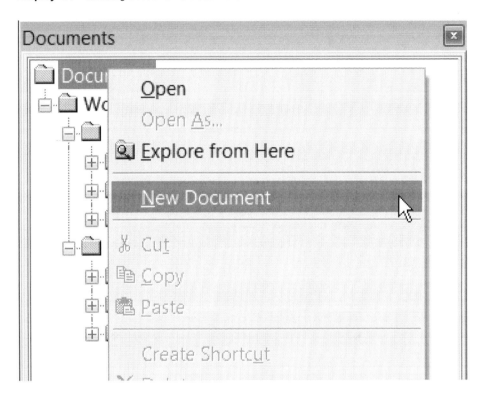

In the New Document window, select a library in which to store the new document, and enter a name for the document. Click **OK**.

Display 4.5 Naming a New ODS Document

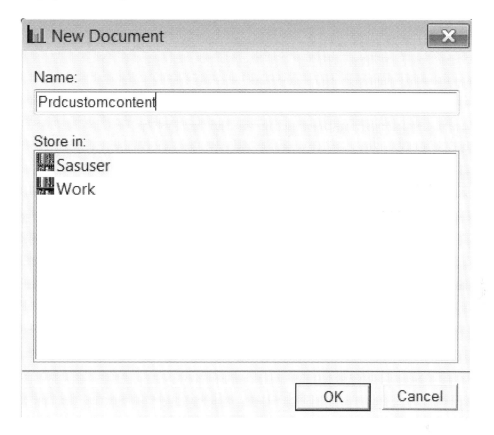

Because you selected the library Sasuser, your document is stored permanently.

Display 4.6 *New Empty Document PrdCustomContent*

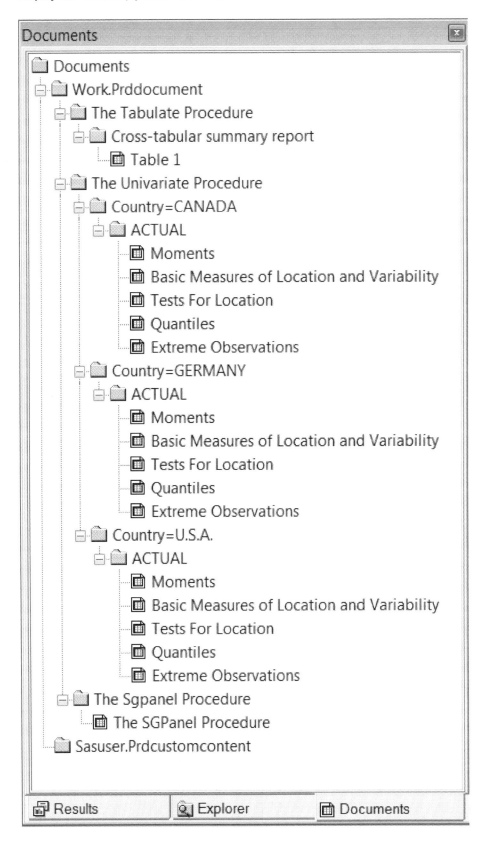

When you create a new document, it is empty. You can load output into the new document by using the Documents window in two ways. You can copy and paste output from the original document (Work.PrdDocument) into the new document (Sasuser.Prdcustomcontent). Or, you can drag and drop output into the new document. You can load individual output objects one by one, as is shown below, or you can add all of the output at once. Then, you can rearrange the output. In the output below under Sasuser.Prdcustomcontent, PROC SGPANEL output is before PROC TABULATE and PROC UNIVARIATE output. Note that you can use PROC DOCUMENT statements to accomplish all these tasks without using the Documents window. For complete documentation on the DOCUMENT procedure, see the Chapter 8, "The DOCUMENT Procedure," in *SAS Output Delivery System: User's Guide*.

Figure 4.1 Adding Output to the Sasuser.Prdcustomcontent Document

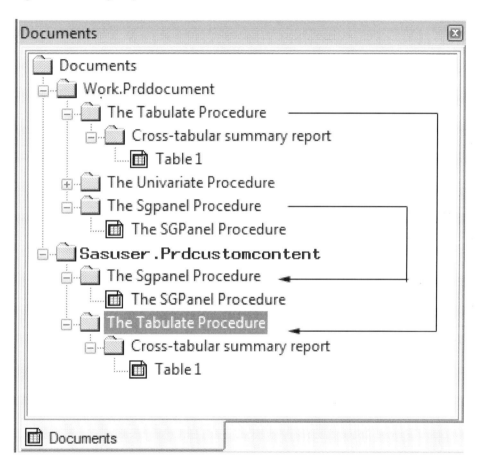

Next, drag and drop the UNIVARIATE procedure output from Work.Prddocument into Sasuser.Prdcustomcontent. All of the output objects that PROC UNIVARIATE creates are copied into the new document.

Figure 4.2 Adding the UNIVARIATE Procedure Output

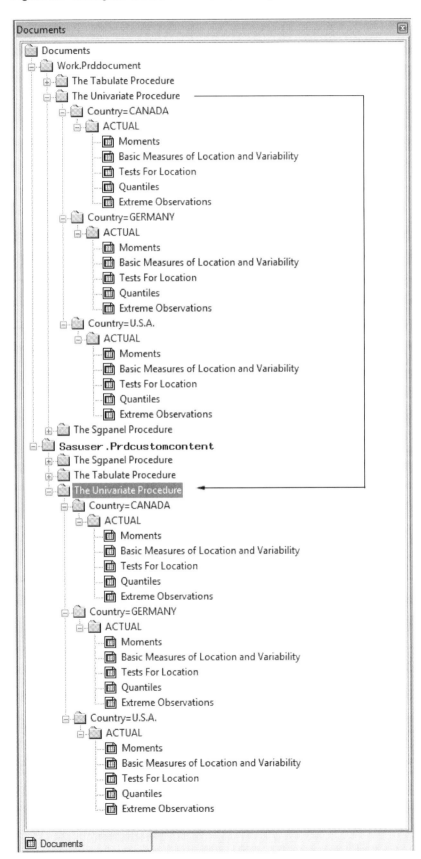

Select the Output Objects **35**

All output objects are in the new document. Delete the PROC UNIVARIATE output objects Basic Measures of Location and Variability and Tests For Location. Right-click on an object, and select **Delete**.

Display 4.7 *Delete Output Objects*

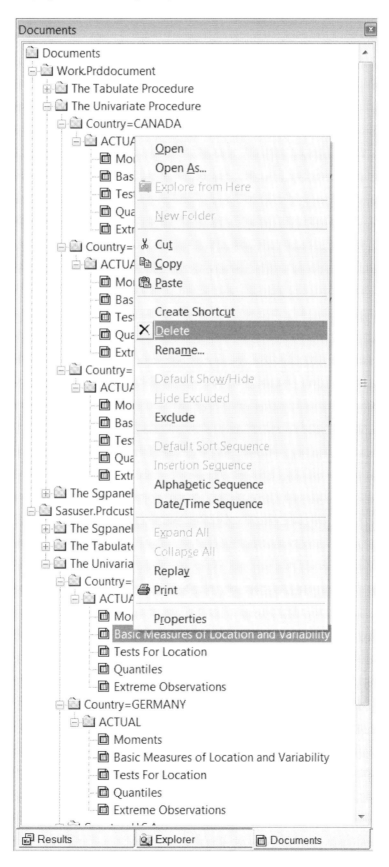

Sasuser.Prdcustomcontent now contains all of the output from PROC SGPANEL and PROC TABULATE. The UNIVARIATE output now consists of only the Moments, Quantiles, and Extreme Observations tables for Canada, Germany, and the United States.

Display 4.8 *Completed New Document*

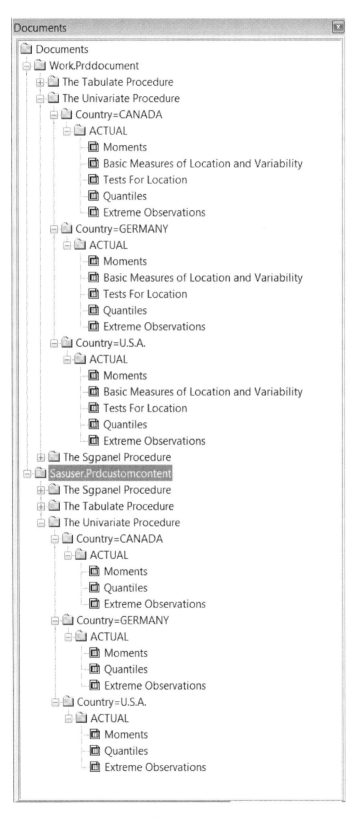

Now that you have the output that you want in the new document, you can re-create the output with only the data that you want shown. Right-click **Sasuser.Prdcustomcontent**, and select **Replay**. This will display your output to any open destinations without rerunning the procedures.

Display 4.9 *Replay the New Document*

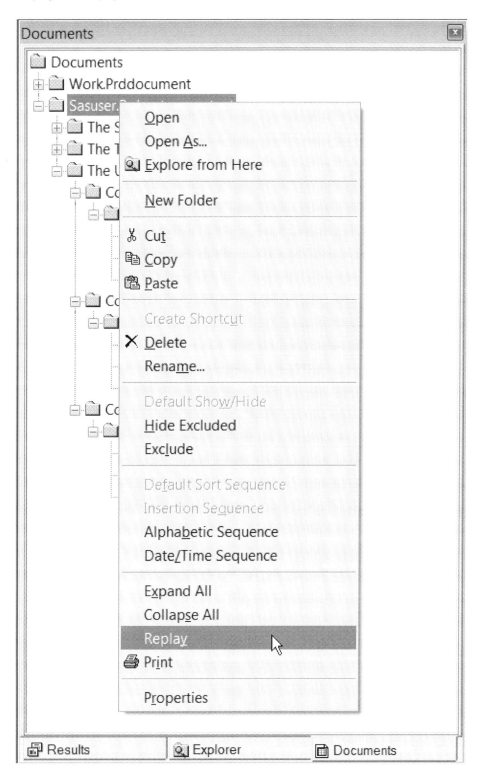

Comparing the Two Methods

If you have your output in an ODS document, you might find it easier to select and exclude output using the Documents window or PROC DOCUMENT. By using PROC DOCUMENT, you do not need to access the data or run the program again. It is easy to rearrange and rename your output objects in the Documents window.

If you do not have your output in an ODS document and you have many output objects to select or exclude, you might want to use the ODS SELECT or ODS EXCLUDE statement. In both of these ODS statements, you can create one line of code that selects or excludes all output objects that you want it to. You can place these ODS statements anywhere in the program.

For More Information

- For complete documentation on ODS statements, see Chapter 6, "Dictionary of ODS Language Statements," in *SAS Output Delivery System: User's Guide*.
- For complete documentation on the DOCUMENT procedure, see Chapter 8, "The DOCUMENT Procedure," in *SAS Output Delivery System: User's Guide*.

Chapter 5
Integrating Output with Popular Business Applications and SAS

About the Tasks That You Will Perform 39
Creating RTF Output 40
Creating PDF Output 42
Creating Enhanced HTML Output 43
Creating Excel Output 45
Combined Program 45
For More Information 48

About the Tasks That You Will Perform

After you select the contents of your report, you can create output for many different business applications. For your report, you are going to create RTF, PDF, HTML, and Excel output. With ODS, it is easy to create output that is formatted for different business applications using ODS destination statements. The ODS statement and the SAS program that it contains form the ODS block.

An ODS block has the following form:

ODS *output-destination 1* <*options(s)*>;

...

ODS *output-destination (n)* <*options(s)*>

<*your SAS program*>

ODS *destination close statement 1*;

...

ODS *destination close statement (n)*

In the ODS block, *output-destination* is the name of a valid ODS destination and *option(s)* are options that are valid for that destination. Your SAS program is inserted between the beginning ODS destination statement and the ODS CLOSE statement.

Most ODS destination statements require the FILE= or BODY= option, in which the name and path of the file that you are generating is specified. It is a good practice to specify one of these options, but it is not always required. By default, if you have not closed and reopened the ODS HTML destination, your output is stored in your temporary directory, unless you specify a different directory with the PATH= option in the ODS HTML statement. After you have opened and closed the ODS HTML destination, your output is stored in your local directory, unless you specify a different directory with the PATH= option.

After you run your program, your HTML, PDF, and Excel output opens in the Results Viewer. RTF output opens in Microsoft Word.

Creating RTF Output

The ODS RTF statement creates RTF output. Suppose you decide that you want only PROC TABULATE output and PROC SGPANEL output in RTF format. To create this output, simply sandwich the ODS RTF statement and ODS RTF CLOSE statement around your program. Use the FILE= option in the ODS RTF statement to specify the name and path for your file. The RTF output will open in Microsoft Word. Because HTML destination is open by default, it is good practice to close the HTML destination if you do not want HTML output. This saves system resources.

```
ods html close;

options nodate nonumber;
proc sort data=sashelp.prdsale out=prdsale;
    by Country;
run;

ods rtf file='your-file-path/RTFPrdsale.rtf';
title 'Actual Product Sales';
title2 '(millions of dollars)';

proc tabulate data=prdsale;
    class region division prodtype;
    classlev region division prodtype;
    var actual;
    keyword all sum;
    keylabel all='Total';
    table (region all)*(division all),
        (prodtype all)*(actual*f=dollar10.) /
        misstext=[label='Missing']
        box=[label='Region by Division and Type'];
 run;

title;
title2;
proc sgpanel data=prdsale ;
    where quarter=1;
    panelby product / novarname;
    vbar region / response=predict;
    vline region / response=actual lineattrs=GraphFit;
    colaxis fitpolicy=thin;
    rowaxis label='Sales';
run;
ods rtf close;
```

Output 5.1 PROC TABULATE Output Viewed in Microsoft Word

Actual Product Sales
(millions of dollars)

Region by Division and Type		Product type		Total
		FURNITURE	OFFICE	
		Actual Sales	Actual Sales	Actual Sales
		Sum	Sum	Sum
Region	Division			
EAST	CONSUMER	$72,570	$108,686	$181,256
	EDUCATION	$73,901	$115,104	$189,005
	Total	$146,471	$223,790	$370,261
WEST	Division			
	CONSUMER	$76,209	$105,020	$181,229
	EDUCATION	$67,945	$110,902	$178,847
	Total	$144,154	$215,922	$360,076
Total	Division			
	CONSUMER	$148,779	$213,706	$362,485
	EDUCATION	$141,846	$226,006	$367,852
	Total	$290,625	$439,712	$730,337

Output 5.2 *PROC SGPANEL Output Viewed in Microsoft Word*

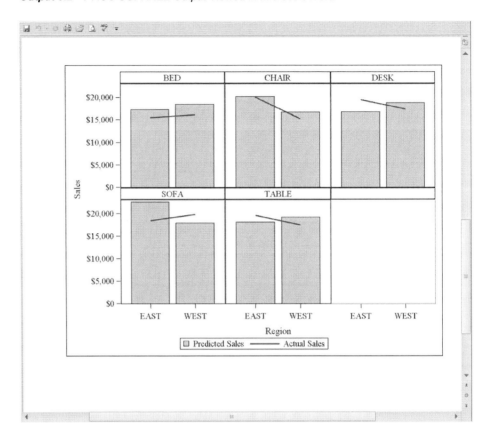

Creating PDF Output

You can generate output that is formatted for Adobe Acrobat software. To create PDF output that contains PROC TABULATE and PROC UNIVARIATE output, sandwich the ODS PDF statement and ODS PDF CLOSE statement around the PROC TABULATE step and PROC UNIVARIATE step. Use the FILE= option to specify the name and path for your file.

```
options nodate nonumber;
proc sort data=sashelp.prdsale out=prdsale;
    by Country;
run;
ods pdf file='your-file-path/PDFPrdsale.pdf';
title 'Actual Product Sales';
title2 '(millions of dollars)';

proc tabulate data=prdsale;
    class region division prodtype;
    classlev region division prodtype;
    var actual;
    keyword all sum;
    keylabel all='Total';
    table (region all)*(division all),
        (prodtype all)*(actual*f=dollar10.) /
        misstext=[label='Missing']
        box=[label='Region by Division and Type'];
```

```
    run;

    title;
    title2;
    ods select ExtremeObs Quantiles Moments;
    proc univariate data=prdsale;
        by Country;
        var actual;
    run;
    ods pdf close;
```

Output 5.3 Default Output Viewed in Adobe Acrobat

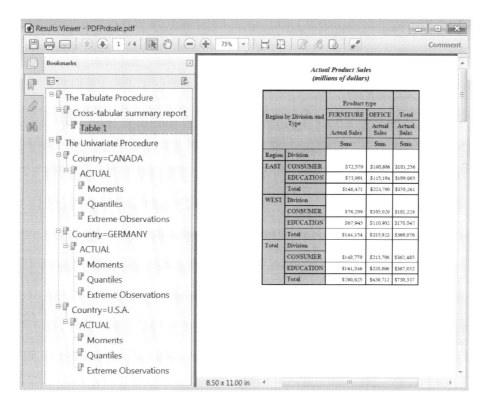

Creating Enhanced HTML Output

By default, if you have not closed and reopened the ODS HTML destination, HTML output is created. You can specify options in the ODS HTML statement to create frame, contents, page, and body files. In the example below, when you run the program, the body file is displayed in the Results Viewer. To view the body file, contents file, and frame file together as a single HTML page, open the frame file from your local directory.

The PATH= option in the ODS HTML statement specifies the directory, folder, or partitioned data set that you are using to store your files. Once you have specified PATH='*your-directory-path*' in the ODS HTML statement, the files html-bodyPrdsale.htm, html-contentsPrdsale.htm, and html-framePrdsale.htm are automatically stored in '*your-directory-path*'.

```
    options nodate nonumber;
    proc sort data=sashelp.prdsale out=prdsale;
```

```
        by Country;
    run;

    ods html path='your-directory-path' body='html-bodyPrdsale.htm'
            contents='html-contentsPrdsale.htm'
            frame='html-framePrdsale.htm';
    title 'Actual Product Sales';
    title2 '(millions of dollars)';
    proc tabulate data=prdsale;
        class region division prodtype;
        classlev region division prodtype;
        var actual;
        keyword all sum;
        keylabel all='Total';
        table (region all)*(division all),
            (prodtype all)*(actual*f=dollar10.) /
            misstext=[label='Missing']
            box=[label='Region by Division and Type'];
    run;

    title;
    title2;
    ods select ExtremeObs Quantiles Moments;
    proc univariate data=prdsale;
        by Country;
        var actual;
    run;
    proc sgpanel data=prdsale;
        where quarter=1;
        panelby product / novarname;
        vbar region / response=predict;
        vline region / response=actual lineattrs=GraphFit;
        colaxis fitpolicy=thin;
        rowaxis label='Sales';
    run;
    ods html close;
```

Output 5.4 *Frame File Created with the ODS HTML Statement Displayed in Internet Explorer*

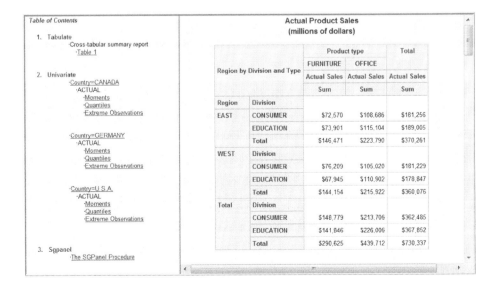

Creating Excel Output

To view output in Excel, you can specify that your original data set be formatted for Excel. Use the ODS TAGSETS.EXCELXP statement to open the Prdsale data set in Excel.

```
ods tagsets.excelxp file='your-file-path/Prdsale.xls';

proc print data=sashelp.prdsale;
run;
ods tagsets.excelxp close;
```

Output 5.5 *Prdsale Data Set Opened in Excel*

Combined Program

Now that you have created output for the various applications that you want to use, you can combine the different programs into a single program. You can create an ODS document to store your output objects. You can then replay your output or modify your output objects without having to resubmit any procedures.

CAUTION:
The following program contains numbered callouts that help explain specific parts of the program. When you copy and paste this program into a SAS code editor, the callouts are copied as well, which results in errors. If you want to copy and paste this program, use Example Code 5.2 on page 47.

Example Code 5.1 *Combined Program with Callouts*

```
options nodate nonumber;
proc sort data=sashelp.prdsale out=prdsale;
    by Country;
run;
```
1 `ods document name=work.prddocument(write);`

46 *Chapter 5 • Integrating Output with Popular Business Applications and SAS*

```
❷ ods html path='your-directory-path'
           body='html-bodyPrdsale.htm'
           contents='html-contentsPrdsale.htm'
           frame='html-framePrdsale.htm';
❸ ods pdf file='your-file-path/PDFPrdsale.pdf' ;
❹ ods rtf file='your-file-path/RTFPrdsale.rtf' ;
   title 'Actual Product Sales';
   title2 '(millions of dollars)';

   proc tabulate data=prdsale;
       class region division prodtype;
       classlev region division prodtype;
       var actual;
       keyword all sum;
       keylabel all='Total';
       table (region all)*(division all),
             (prodtype all)*(actual*f=dollar10.) /
             misstext=[label='Missing']
             box=[label='Region by Division and Type'];
   run;

   title;
   title2;
❺ ods rtf exclude all;
❻ ods select ExtremeObs Quantiles Moments;

   proc univariate data=prdsale;
   by Country;
   var actual;
   run;

❼ ods rtf select all;
❽ ods pdf select none;

   title 'Sales Figures for First Quarter by Product';
   proc sgpanel data=prdsale;
       where quarter=1;
       panelby product / novarname;
       vbar region / response=predict;
       vline region / response=actual lineattrs=GraphFit;
       colaxis fitpolicy=thin;
       rowaxis label='Sales';
   run;

❾ ods exclude PRINT;
❿ ods tagsets.excelxp file=your-file-path.'Prdsale.xls';
⓫ ods tagsets.excelxp select PRINT;
   proc print data=sashelp.prdsale;
   run;
⓬ ods _all_ close;
⓭ ods html;
```

1   The ODS DOCUMENT statement creates the document Work.PrdDocument. Work.PrdDocument stores all of the output generated between the opening ODS DOCUMENT statement and the ODS _ALL_ CLOSE statement.

2   The ODS HTML statement specifies the names and paths for the body, contents, and frame files.

3   The ODS PDF statement with the FILE= option opens the ODS PDF destination (which is a member of the PRINTER family of destinations). It specifies the name and path for the PDF output file.

4   The ODS RTF statement with the FILE= option opens the ODS RTF destination. It specifies the name and path for the RTF output file.

5   The ODS RTF statement with the EXCLUDE ALL option excludes all of the output objects from the following PROC UNIVARIATE output.

6   The ODS SELECT statement specifies that the output objects ExtremeObs, Quantiles, and Moments be sent to all open destinations that do not specifically exclude PROC UNIVARIATE output with the EXCLUDE option, such as the previous ODS RTF statement. The ODS statement with the SELECT or EXCLUDE option must be specified after the opening ODS statement.

7   The ODS RTF statement with the SELECT ALL option selects all of the output objects from the following PROC SGPANEL output. It sends the output objects to the ODS RTF destination. The ODS statement with the SELECT or EXCLUDE option must be specified after the opening ODS statement.

8   The ODS PDF statement with the SELECT NONE option selects none of the output objects from the following PROC SGPANEL output. The ODS statement with the SELECT or EXCLUDE option must be specified after the opening ODS statement.

9   The ODS EXCLUDE statement excludes the output object named Print from all open destinations that do not specifically select the Print output object with the SELECT option.

10 The ODS TAGSETS.EXCELXP statement with the FILE= option opens the TAGSETS.EXCELXP destination (which is a member of the MARKUP family of destinations). It specifies the name and path for the XLS output file. You can use the .XML extension instead of the EXCELXP extension. Beginning in Excel 2007, using the XLS extension will invoke a dialog box because of the new security feature that matches the content with the extension.

11 The ODS TAGSETS.EXCELXP statement with the SELECT option selects the output object named Print.

12 The ODS _ALL_ CLOSE statement closes all of the open destinations. This statement is useful when you have multiple destinations open at the same time.

13 Because the ODS _ALL_ CLOSE statement closes all open destinations, it is a good practice to open the ODS HTML destination again at the end of your program. If all of the destinations are closed, you get the following warning in the SAS Log: **WARNING: No output destinations active**.

Example Code 5.2 *Combined Program without Callouts*

```
options nodate nonumber;
proc sort data=sashelp.prdsale out=prdsale;
    by Country;
run;
ods document name=work.prddocument(write);

ods html path='your-directory-path'
        body='html-bodyPrdsale.htm'
        contents='html-contentsPrdsale.htm'
        frame='html-framePrdsale.htm';
```

```
ods pdf file='your-file-path/PDFPrdsale.pdf' ;
ods rtf file='your-file-path/RTFPrdsale.rtf' ;
title 'Actual Product Sales';
title2 '(millions of dollars)';

proc tabulate data=prdsale;
    class region division prodtype;
    classlev region division prodtype;
    var actual;
    keyword all sum;
    keylabel all='Total';
    table (region all)*(division all),
        (prodtype all)*(actual*f=dollar10.) /
        misstext=[label='Missing']
        box=[label='Region by Division and Type'];
run;

title;
title2;
ods rtf exclude all;
ods select ExtremeObs Quantiles Moments;

proc univariate data=prdsale;
by Country;
var actual;
run;

ods rtf select all;
ods pdf select none;

title 'Sales Figures for First Quarter by Product';
proc sgpanel data=prdsale;
    where quarter=1;
    panelby product / novarname;
    vbar region / response=predict;
    vline region / response=actual lineattrs=GraphFit;
    colaxis fitpolicy=thin;
    rowaxis label='Sales';
run;

ods exclude PRINT;
ods tagsets.excelxp file=your-file-path.'Prdsale.xls';
ods tagsets.excelxp select PRINT;
proc print data=sashelp.prdsale;
run;
ods _all_ close;
ods html;
```

For More Information

- For complete documentation on ODS statements, see the Chapter 6, "Dictionary of ODS Language Statements," in *SAS Output Delivery System: User's Guide*.

Chapter 6
Customizing the Presentation of a Report

About the Tasks That You Will Perform 49

Customized RTF Output 51

Customized PROC TABULATE Output 52

Customized PDF Output 54

Customized HTML Output 56

Customized Excel Output 57

Combined Program 59

For More Information 63

About the Tasks That You Will Perform

Once you have specified the ODS statements that you need to generate the appropriate output type, you can add options to those statements to customize the presentation of your output. The quickest way to make a change to all of your output at once is to specify the STYLE= option in the ODS destination statement.

You can view all styles that are provided by SAS in the Templates window. To open the Templates window, do one of the following:

- Select **Results** in the Results window, and then select **View** ⇨ **Templates** from the taskbar.

- Right-click **Results** in the Results window, and then select **Templates**.

- Issue the following command on the command line in the SAS windowing environment:

  ```
  odstemplates
  ```

To view the styles provided by SAS, select **Templates** ⇨ **Sashelp.Tmplmst** ⇨ **Styles**.

Output 6.1 SAS Styles

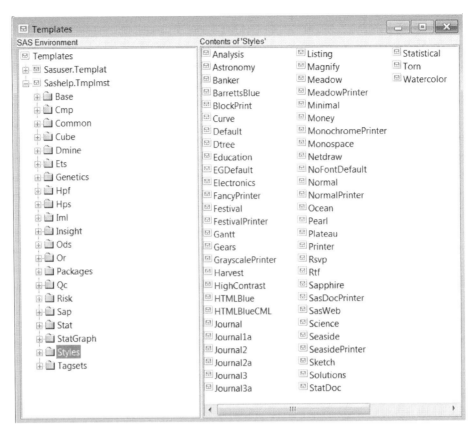

By default (for DMS), ODS uses the HTMLBlue style for HTML output, the RTF style for RTF output, and the default Printer style for PDF output. The default style for each ODS destination can be viewed in the SAS Registry Editor. For the following examples, the Science style is used to ensure that the look of the presentation is consistent. To view the source of the Science style in the Template Browser window, do one of the following:

- Double-click the **Science** style.
- Right-click **Science**, and then select **Open**.

Display 6.1 Source of Science Style Viewed in the Template Browser Window

You can customize specific areas of some procedure output by specifying the STYLE= option in specific procedure statements. With the STYLE= option, you can change the attributes of cells. Changes include aligning text, adding URLs, changing rules, and changing the size, width, and color of a font.

The STYLE= option is specified in slightly different ways for individual procedure statements. However, the general form of the STYLE= option for a procedure statement is:

STYLE=<*style-attribute-name=style-attribute-value*>

The procedures that allow direct style control are PROC PRINT, PROC REPORT, and PROC TABULATE. For documentation on these procedures, see the *Base SAS Procedures Guide*.

Customized RTF Output

To quickly change the look of your RTF output, specify the STYLE= and STARTPAGE= option in the ODS RTF statement. The STYLE= option with the Science style specified tells ODS to use the Science style instead of the default style for RTF output. Because the HTML destination is open by default, it is good practice to close the HTML destination if you do not want HTML output. This saves system resources.

```
ods html close;
options nodate nonumber;
proc sort data=sashelp.prdsale out=prdsale;
    by Country;
run;
ods rtf file='your-file-path/RTFPrdsaleCustom.rtf' style=Science;

title 'Actual Product Sales';
title2 '(millions of dollars)';

proc tabulate data=prdsale;
    class region division prodtype;
    classlev region division prodtype;
    var actual;
    keyword all sum;
    keylabel all='Total';
    table (region all)*(division all),
          (prodtype all)*(actual*f=dollar10.) /
          misstext=[label='Missing']
          box=[label='Region by Division and Type'];
 run;

title2;
title;

proc sgpanel data=prdsale ;
    where quarter=1;
    panelby product / novarname;
    vbar region / response=predict;
    vline region / response=actual lineattrs=GraphFit;
    colaxis fitpolicy=thin;
```

```
        rowaxis label='Sales';
run;
ods rtf close;
```

Even though you did not specify the STYLE= option in the PROC SGPANEL statement, the `style=Science` option in the ODS RTF statement applies the style for PROC SGPANEL, too.

Output 6.2 Customized RTF Output Viewed in Microsoft Word

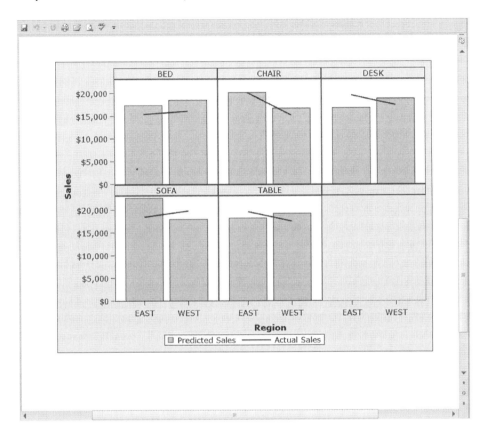

Customized PROC TABULATE Output

You can use ODS options in Base SAS reporting procedures, PROC PRINT, PROC REPORT, and PROC TABULATE. In each of these procedures, you can specifically specify options in individual statements. This enables you to make changes in sections of output without changing the default style of all of the output. For example, you can customize specific sections of PROC TABULATE output by specifying the STYLE= option in specific statements within the procedure. Because the HTML destination is open by default, it is good practice to close the HTML destination if you do not want HTML output. This saves system resources.

```
ods html close;
options nodate nonumber;
proc sort data=sashelp.prdsale out=prdsale;
    by Country;
run;
ods rtf file='your-file-path/RTFPrdsaleCustom.rtf' style=Science;
```

```
title 'Actual Product Sales';
title2 '(millions of dollars)';

proc tabulate data=prdsale ❶style=[fontweight=bold];
    class region division prodtype /❷ style=[textalign=center];
    classlev region division prodtype / ❸style=[textalign=left];
    var actual / ❹style=[fontsize=3];
    keyword all sum;
    keylabel all='Total';
    table (region all)*(division all*❺[style=[backgroundcolor=yellow]]),
          (prodtype all)*(actual*f=dollar10.) /
          style=[bordercolor=blue] box=[label='Region by Division and Type'
          ❻style=[fontstyle=italic]];
run;

title2;
title;

proc sgpanel data=prdsale;
    where quarter=1;
    panelby product / novarname;
    vbar region / response=predict;
    vline region / response=actual lineattrs=GraphFit;
    colaxis fitpolicy=thin;
    rowaxis label='Sales';
run;
ods rtf close;
```

1   The STYLE= option specified in the PROC TABULATE statement changes all of the font to bold for all of the data cells.

2   The STYLE= option specified in the CLASS statement centers the CLASS variable name headings.

3   The STYLE= option specified in the CLASSLEV statement left-justifies the CLASS variable level value headings.

4   The STYLE= option specified in the VAR statement changes the font size of analysis variable name headings to 3.

5   The first STYLE= option specified in the TABLE statement changes the background color of the cells containing the sum totals of REGION and DIVISION to yellow.

6   The second STYLE= option specified in the TABLE statement italicizes the font of the label of the empty box above the row titles.

Because the STYLE= option is specified in the ODS RTF statement, PROC TABULATE output uses the Science style and the specific style overrides specified in individual statements.

Output 6.3 *Customized PROC TABULATE Output Viewed in Microsoft Word*

Actual Product Sales (millions of dollars)

Region by Division and Type		Product type		Total
		FURNITURE	OFFICE	
		Actual Sales	Actual Sales	Actual Sales
		Sum	Sum	Sum
Region	Division			
EAST	CONSUMER	$72,570	$108,686	$181,256
	EDUCATION	$73,901	$115,104	$189,005
	Total	$146,471	$223,790	$370,261
WEST	Division			
	CONSUMER	$76,209	$105,020	$181,229
	EDUCATION	$67,945	$110,902	$178,847
	Total	$144,154	$215,922	$360,076
Total	Division			
	CONSUMER	$148,779	$213,706	$362,485
	EDUCATION	$141,846	$226,006	$367,852
	Total	$290,625	$439,712	$730,337

Customized PDF Output

To quickly change the look of your PDF output, specify the STYLE=, CONTENTS=, and PDFTOC= options in the ODS PDF statement. The STYLE= option with the Science style specified tells ODS to use the Science style for all PDF output. The CONTENTS= option with YES specified creates a table of contents for your PDF file. The PDFTOC= option with 2 specified expands the table of contents to two levels.

```
options nodate nonumber orientation=portrait;
proc sort data=sashelp.prdsale out=prdsale;
    by Country;
run;

ods pdf file='your-file-path/PDFPrdsaleCustom.pdf' contents=yes pdftoc=2 style=S
title 'Actual Product Sales';
title2 '(millions of dollars)';
```

```
proc tabulate data=prdsale style=[fontweight=bold];
    class region division prodtype / style=[textalign=center];
    classlev region division prodtype / style=[textalign=left];
    var actual / style=[fontsize=3];
    keyword all sum;
    keylabel all='Total';
    table (region all)*(division all*[style=[backgroundcolor=yellow]]),
          (prodtype all)*(actual*f=dollar10.) /
          style=[bordercolor=blue] style=[fontstyle=italic]];
run;

title;
title2;

proc univariate data=prdsale;
    by Country;
    var actual;
run;
ods pdf close;
```

Output 6.4 *Customized Table of Contents Created and Viewed in Adobe Acrobat*

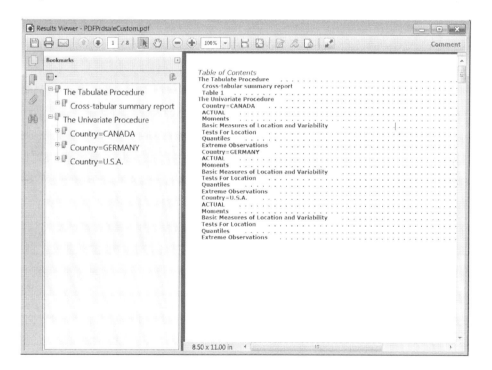

Output 6.5 *Customized PDF Output Viewed in Adobe Acrobat*

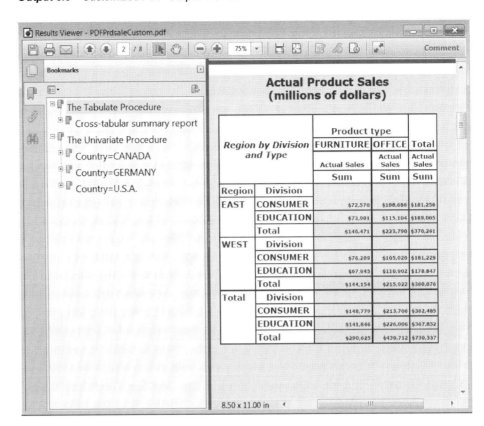

Customized HTML Output

The STYLE= option with the Science style specified in the ODS HTML statement tells ODS to use the Science style for all HTML output.

```
options nodate nonumber;
proc sort data=sashelp.prdsale out=prdsale;
    by Country;
run;
ods html path='your-directory-path'
        body='bodyPrdsale.htm'
        contents='contentsPrdsale.htm'
        frame='framePrdsale.htm'
        style=Science;
title 'Actual Product Sales';
title2 '(millions of dollars)';

proc tabulate data=prdsale style=[fontweight=bold];
    class region division prodtype / style=[textalign=center];
    classlev region division prodtype / style=[textalign=left];
    var actual / style=[fontsize=3];
    keyword all sum;
    keylabel all='Total';
    table (region all)*(division all*[style=[backgroundcolor=yellow]]),
        (prodtype all)*(actual*f=dollar10.) /
        style=[bordercolor=blue]
```

```
            misstext=[label='Missing' style=[fontweight=light]]
            box=[label='Region by Division and Type'
            style=[fontstyle=italic]];
   run;

   title;
   title2;

   ods select ExtremeObs Quantiles Moments;

   proc univariate data=prdsale;
       by Country;
       var actual;
   run;
   proc sgpanel data=prdsale;
       where quarter=1;
       panelby product / novarname;
       vbar region / response=predict;
       vline region / response=actual lineattrs=GraphFit;
       colaxis fitpolicy=thin;
       rowaxis label='Sales';
   run;
   ods html close;
```

Output 6.6 *Customized Frame File Created and Viewed in Internet Explorer*

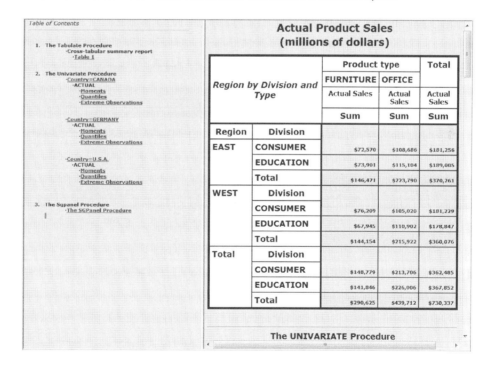

Customized Excel Output

The STYLE= option with the Science style specified in the ODS TAGSETS.EXCELXP statement tells ODS to use the Science style for all Excel output. Specifying the OPTIONS suboption (DOC="HELP") prints Help for the ODS TAGSETS.EXCELXP statement suboptions to the SAS log file. All tagsets have instream help. For more information about the DOC= suboption, see the ODS Tagset statement in Chapter 6, "Dictionary of ODS Language Statements," in *SAS Output Delivery System: User's Guide*.

```
ods tagsets.excelxp file='your-file-path/Prdsale.xls' style=Science options
(doc="help");

proc print data=sashelp.prdsale;
run;
ods tagsets.excelxp close;
```

Output 6.7 *Customized Excel Output*

Obs	ACTUAL	PREDICT	COUNTRY	REGION
1	$925.00	$850.00	CANADA	EAST
2	$999.00	$297.00	CANADA	EAST
3	$608.00	$846.00	CANADA	EAST
4	$642.00	$533.00	CANADA	EAST
5	$656.00	$646.00	CANADA	EAST
6	$948.00	$486.00	CANADA	EAST
7	$612.00	$717.00	CANADA	EAST
8	$114.00	$564.00	CANADA	EAST
9	$685.00	$230.00	CANADA	EAST
10	$657.00	$494.00	CANADA	EAST
11	$608.00	$903.00	CANADA	EAST
12	$353.00	$266.00	CANADA	EAST
13	$107.00	$190.00	CANADA	EAST
14	$354.00	$139.00	CANADA	EAST
15	$101.00	$217.00	CANADA	EAST
16	$553.00	$560.00	CANADA	EAST
17	$877.00	$148.00	CANADA	EAST
18	$431.00	$762.00	CANADA	EAST

Display 6.2 *Help for the EXCELXP Tagset*

```
Log - (Untitled)
50    ods tagsets.excelxp file='Prdsale.xls' style=Science options (doc="help");
NOTE: Writing TAGSETS.EXCELXP Body file: Prdsale.xls
============================================================================
The EXCELXP Tagset Help Text.

This Tagset/Destination creates Microsoft's spreadsheetML XML.
It is used specifically for importing data into Excel.

Each table will be placed in its own worksheet within a workbook.
This destination supports ODS styles, traffic lighting, and custom formats.

Numbers, Currency and percentages are correctly detected and displayed.
Custom formats can be given by supplying a style override on the tagattr
style element.

By default, titles and footnotes are part of the spreadsheet, but are part
of the header and footer.

Also by default, printing will be in 'Portrait'.
The orientation can be changed to landscape.

The specification for this xml is here.
http://msdn.microsoft.com/library/default.asp?url=/library/en-us/dnexcl2k2/html/odc_xlsmlin

See Also:
http://support.sas.com/rnd/base/topics/odsmarkup/
http://support.sas.com/rnd/papers/index.html#excelxml

Sample usage:

ods tagsets.excelxp file='test.xml' contents='index.xml' data='test.ini' options(doc='Help'

ods tagsets.excelxp options(doc='Quick');

ods tagsets.excelxp options(embedded_titles='No' Orientation='Landscape');

============================================================================
Long descriptions of the supported options

Doc:  No default value.
      Help: Displays introductory text and available options in full detail.
      Quick: Displays introductory text and an alphabetical list of options,
             their current value, and a short description
```

Combined Program

Now that you have created output for the various applications that you want to use, you can combine the different programs into a single program. You can create an ODS document to store your output objects. You can then replay your output or modify your output objects without having to resubmit any procedures.

CAUTION:

The following program contains numbered callouts that help explain specific parts of the program. When you copy and paste this program into a SAS code editor, the callouts are copied as well, which results in errors. If you want to copy and paste this program, use Example Code 6.2 on page 62.

Example Code 6.1 *Combined Program with Callouts*

```
proc sort data=sashelp.prdsale out=prdsale;
    by Country;
run;
options nodate nonumber;
❶ods document name=work.prddocument(write);

❷ods html path='your-directory-path' body='bodyPrdsale.htm'
        contents='contentsPrdsale.htm'
        frame='framePrdsale.htm'
        style=Science;

    ❸ods rtf file='your-file-path/RTFPrdsaleCustom.rtf' style=Science;
```

```
          4 ods pdf file='your-file-path/PDFPrdsaleCustom.pdf'
               contents=yes pdftoc=2 style=Science;
title 'Actual Product Sales';
title2 '(millions of dollars)';

proc tabulate data=prdsale 5 style=[fontweight=bold];
    class region division prodtype / 6 style=[textalign=center];
    classlev region division prodtype / 7 style=[textalign=left];
    var actual / 8 style=[fontsize=3];
    keyword all sum;
    keylabel all='Total';
    table (region all)*(division all* 9 [style=[backgroundcolor=yellow]]),
          (prodtype all)*(actual*f=dollar10.) /
          style=[bordercolor=blue] box=[label='Region by Division and Type'
          10 style=[fontstyle=italic]];
run;
title;
title2;

11 ods rtf exclude all;
12 ods select ExtremeObs Quantiles Moments;

proc univariate data=prdsale;
by Country;
var actual;
run;

13 ods rtf select all;
14 ods pdf select none;

title 'Sales Figures for First Quarter by Product';
proc sgpanel data=prdsale;
    where quarter=1;
    panelby product / novarname;
    vbar region / response=predict;
    vline region / response=actual lineattrs=GraphFit;
    colaxis fitpolicy=thin;
    rowaxis label='Sales';
run;

15 ods exclude PRINT;
16 ods tagsets.excelxp file='your-file-path/Prdsale.xls' style=Science;
17 ods tagsets.excelxp select PRINT;
proc print data=sashelp.prdsale;
run;
18 ods _all_ close;
19 ods html;
```

1. The ODS DOCUMENT statement creates the document Work.PrdDocument. Work.PrdDocument stores all of the output generated between the opening ODS DOCUMENT statement and the ODS _ALL_ CLOSE statement.

2. The ODS HTML statement specifies the names and paths for the body, contents, frame, and page files. The STYLE=SCIENCE option applies the Science style to all HTML output.

3   The ODS RTF statement with the FILE= option opens the ODS RTF destination. It specifies the name and path for the RTF output file. The STYLE=SCIENCE option applies the Science style to all RTF output. The STARTPAGE=YES option specifies to put each procedure's output on a new page.

4   The ODS PDF statement with the FILE= option opens the ODS PDF destination. It specifies the name and path for the PDF output file. The CONTENTS=YES option creates a table of contents for your PDF file. The PDFTOC=2 option specifies that the table of contents is expanded to two levels.

5   The STYLE= option specified in the PROC TABULATE statement changes all of the font to bold.

6   The STYLE= option specified in the CLASS statement centers the CLASS variable name headings.

7   The STYLE= option specified in the CLASSLEV statement left-justifies the CLASS variable level value headings.

8   The STYLE= option specified in the VAR statement changes the font size of ANALYSIS variable name headings to 3 point.

9   The first STYLE= option specified in the TABLE statement changes the background color of the cells containing the sum totals of REGION and DIVISION to yellow.

10 The second STYLE= option specified in the TABLE statement italicizes the font of the label of the empty box above the row titles.

11 The ODS RTF statement with the EXCLUDE ALL option excludes all of the output objects from the following PROC UNIVARIATE output.

12 The ODS SELECT statement specifies that the output objects ExtremeObs, Quantiles, and Moments be sent to all open destinations that do not specifically exclude PROC UNIVARIATE output with the EXCLUDE option, such as the previous ODS RTF statement. The ODS statement with the SELECT or EXCLUDE option must be specified after the opening ODS statement.

13 The ODS RTF statement with the SELECT ALL option selects all of the output objects from the following PROC SGPANEL output. It sends the output objects to the ODS RTF destination. The ODS statement with the SELECT or EXCLUDE option must be specified after the opening ODS statement.

14 The ODS PDF statement with the SELECT NONE option selects none of the output objects from the following PROC SGPANEL output. The ODS statement with the SELECT or EXCLUDE option must be specified after the opening ODS statement.

15 The ODS EXCLUDE statement excludes the output object named Print from all open destinations that do not specifically select the Print output object with the SELECT option.

16 The ODS TAGSETS.EXCELXP statement with the FILE= option opens the TAGSETS.EXCELXP destination (which is a member of the MARKUP family of destinations). It specifies the name and path for the XLS output file. You can use the .XML extension instead of the EXCELXP extension. Beginning in Excel 2007, using the XLS extension will invoke a dialog box because of a new security feature that matches the content with the extension. The STYLE=SCIENCE option applies the Science style to all TAGSETS.EXCELXP output.

17 The ODS TAGSETS.EXCELXP statement with the SELECT option selects the output object named Print.

18 The ODS _ALL_ CLOSE statement closes all of the open destinations. This statement is useful when you have multiple destinations open at the same time.

19 Because the ODS _ALL_ CLOSE statement closes all open destinations, it is a good practice to open the ODS HTML destination again at the end of your program. If all of the destinations are closed, you get the following warning in the SAS Log: **WARNING: No output destinations active.**

Example Code 6.2 *Combined Program without Callouts*

```
proc sort data=sashelp.prdsale out=prdsale;
    by Country;
run;
options nodate nonumber;
ods document name=work.prddocument(write);

ods html path='your-directory-path' body='bodyPrdsale.htm'
        contents='contentsPrdsale.htm'
        frame='framePrdsale.htm'
        style=Science;

        ods rtf file='your-file-path/RTFPrdsaleCustom.rtf' style=Science;
        ods pdf file='your-file-path/PDFPrdsaleCustom.pdf'
        contents=yes pdftoc=2 style=Science;
title 'Actual Product Sales';
title2 '(millions of dollars)';

proc tabulate data=prdsale style=[fontweight=bold];
    class region division prodtype / style=[textalign=center];
    classlev region division prodtype / style=[textalign=left];
    var actual / style=[fontsize=3];
    keyword all sum;
    keylabel all='Total';
    table (region all)*(division all*[style=[backgroundcolor=yellow]]),
          (prodtype all)*(actual*f=dollar10.) /
          style=[bordercolor=blue] box=[label='Region by Division and Type'
          style=[fontstyle=italic]];
run;
title;
title2;

ods rtf exclude all;
ods select ExtremeObs Quantiles Moments;

proc univariate data=prdsale;
by Country;
var actual;
run;

ods rtf select all;
ods pdf select none;

title 'Sales Figures for First Quarter by Product';
proc sgpanel data=prdsale;
    where quarter=1;
    panelby product / novarname;
    vbar region / response=predict;
    vline region / response=actual lineattrs=GraphFit;
    colaxis fitpolicy=thin;
    rowaxis label='Sales';
```

```
run;

ods exclude PRINT;
ods tagsets.excelxp file='your-file-path/Prdsale.xls' style=Science;
ods tagsets.excelxp select PRINT;
proc print data=sashelp.prdsale;
run;
ods _all_ close;
ods html;
```

For More Information

- For conceptual information about ODS statements, see Chapter 5, "Introduction to ODS Language Statements," in *SAS Output Delivery System: User's Guide*.

- For syntax and usage information about ODS statements, see the Chapter 6, "Dictionary of ODS Language Statements," in *SAS Output Delivery System: User's Guide*.

- For conceptual information about styles, see the Chapter 3, "Output Delivery System: Basic Concepts," in *SAS Output Delivery System: User's Guide*.

- For complete documentation on the Base SAS reporting procedures, see the *Base SAS Procedures Guide*.

- For information about ODS styles, see the Chapter 13, "TEMPLATE Procedure: Creating a Style Template (Definition)," in *SAS Output Delivery System: User's Guide*.

- For specific information about specifying the STYLE= option in PROC TABULATE statements, see the Chapter 53, "TABULATE Procedure," in *Base SAS Procedures Guide*.

Chapter 7
Next Steps: A Quick Look at Advanced Features

Working with the TEMPLATE Procedure 65
    Introduction to the TEMPLATE Procedure 65
    What Can You Do with the TEMPLATE Procedure? 66

Working with ODS Graphics 70

Advanced Features of the DOCUMENT Procedure 74
    Overview 74
    Working with the DOCUMENT Procedure 75

ODS and the DATA Step 78

Where to Go from Here 80

Working with the TEMPLATE Procedure

Introduction to the TEMPLATE Procedure

The TEMPLATE procedure enables you to customize the appearance of your SAS output. For example, you can create, extend, or modify existing templates for various types of output, such as the following:

- styles
- tables
- crosstabulation tables
- columns
- headers
- footers
- tagsets
- ODS Graphics

ODS uses these templates to produce formatted output.

You can use the TEMPLATE procedure to navigate and manage the templates stored in template stores. Here are some tasks that you can do with PROC TEMPLATE:

- Edit an existing template.
- Create links to an existing template.

- Change the location where you write new templates.
- Search for existing templates.
- View the source code of a template.

What Can You Do with the TEMPLATE Procedure?

Modify a Table Template That a SAS Procedure Uses

The following output shows the use of a customized table template for the Moments output object using PROC UNIVARIATE. The program that creates the modified table template does the following:

- Creates and edits a copy of the default table template.
- Edits a header in the table template.
- Sets column attributes to enhance the appearance of the HTML output.

To view the code that creates the following output, see the "Example 1: Editing a Table Template That a SAS Procedure Uses" in Chapter 14 of *SAS Output Delivery System: User's Guide*.

Output 7.1 Default Moments Table

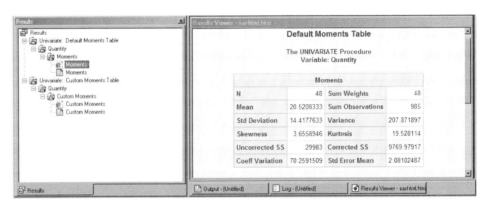

Output 7.2 Customized Moments Table (Customized HTML Output from PROC UNIVARIATE and Viewed with Microsoft Internet Explorer)

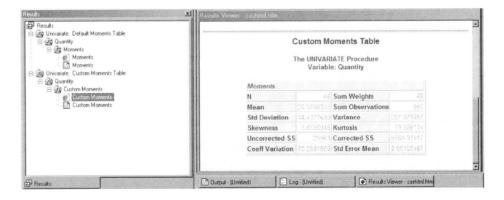

Modify a Style

When you are working with styles, you are more likely to modify a style that SAS provides than create a completely new style. The following output uses the

Styles.HTMLBlue template that SAS provides and includes changes made to the style to customize the output's appearance. To view the code that creates this output, see the "Example 3: Using the CLASS Statement" in Chapter 13 of *SAS Output Delivery System: User's Guide*.

In the contents file, changes to the style are to the following:

- The text of the header and the text that identifies the procedure that produced the output.
- The colors for some of the text.
- The font size for some of the text.
- The spacing in the entries in the table of contents.

In the body file, changes to the style are to the following:

- Two of the colors in the color list. One of these colors is used as the foreground color for the table of contents, the BY line, and column headings. The other color is used as the foreground color for many parts of the body file, including SAS titles and footnotes.
- The font size for titles and footnotes.
- The font style for headers.
- The presentation of the data in the table by changing attributes like cell spacing, rules, and border width.

Display 7.1 *HTML Output (Viewed with Microsoft Internet Explorer)*

Contents	Energy Expenditures for Each Region (millions of dollars)		
1. Print	Division=Middle Atlantic		
·Division=Middle Atlantic	State	Type	Expenditures
·Data Set WORK.ENERGY	NY	Residential Customers	8,786
·Division=Mountain	NY	Business Customers	7,825
·Data Set WORK.ENERGY	NJ	Residential Customers	4,115
	NJ	Business Customers	3,558
	PA	Residential Customers	6,478
	PA	Business Customers	3,695
	Division=Mountain		
	State	Type	Expenditures
	MT	Residential Customers	322
	MT	Business Customers	232
	ID	Residential Customers	392
	ID	Business Customers	298
	WY	Residential Customers	194
	WY	Business Customers	184
	CO	Residential Customers	1,215
	CO	Business Customers	1,173

Create Your Own Tagset

Tagsets are used to create custom markup. You can create your own tagsets, extend existing tagsets, or modify a tagset that SAS provides. The following display shows the results from a new tagset named TAGSET.MYTAGS.

To see the customized MYTAGS.CHTML tagset, view the source by selecting **View** ⇨ **Source** from your Web browser's toolbar.

Display 7.2 MYTAGS.CHTML Output (Viewed with Microsoft Internet Explorer)

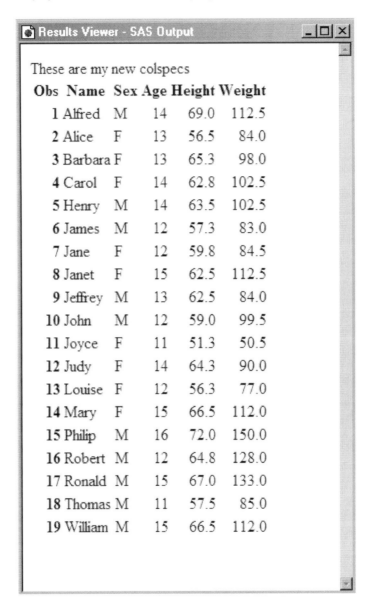

Create a Template-Based Graph

StatGraph templates create output called ODS Graphics. For complete information, see the *SAS Graph Template Language: User's Guide*.

The following code creates the StatGraph template MyGraphs.Regplot, which creates the following graph:

```
proc template;
define statgraph mygraphs.regplot;
begingraph;
```

```
        entrytitle "Regression Plot";
        layout overlay;
          modelband "mean";
          scatterplot x=height y=weight;
          regressionplot x=height y=weight / clm="mean";
        endlayout;
      endgraph;
    end;
    run;

    ods listing style=analysis;
    ods graphics / reset imagename="reg" width=500px;

    proc sgrender data=sashelp.class template=mygraphs.regplot;
    run;
```

The following display shows a scatterplot with an overlaid regression line and confidence limits for the mean of the HEIGHT and WEIGHT variables of a data set.

Display 7.3 *Scatterplot Created with a StatGraph Template*

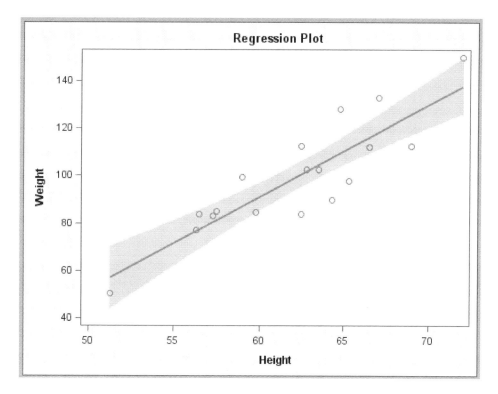

Modify a Crosstabulation Table

The TEMPLATE procedure enables you to customize the appearance of crosstabulation (contingency) tables that are created with the FREQ procedure. By default, crosstabulation tables are formatted based on the CrossTabFreqs template that SAS provides. However, you can create a customized CrossTabFreqs template using the TEMPLATE procedure with the DEFINE CROSSTABS statement. To view the SAS code that creates this output, see the "Example 2: Creating a Crosstabulation Table Template with a Customized Legend" in Chapter 11 of *SAS Output Delivery System: User's Guide*.

The following output shows the use of a customized CrossTabFreqs template for the CrossTabFreqs table. The program that creates the customized CrossTabFreqs template does the following:

- Modifies table regions.
- Customizes legend text.
- Modifies headers and footers.
- Modifies variable labels used in headers.
- Customizes styles for cellvalues.

Display 7.4 Customized CrossTabFreqs Template for the CrossTabFreqs Table

City Government Form by Number of Meetings Scheduled

The FREQ Procedure

Frequency Percent Row Percent Column Percent	City Government Form	Number of Meetings Scheduled						
		?	Not Known	100 or Less	101-200	201-300	Over 300	Total
	?	0	0	0	1	0	0	
	Not Applicable	0	10	0	0	0	0	
	Council Manager	0	0	47 12.30 22.27 55.95	63 16.49 29.86 58.88	49 12.83 23.22 62.03	52 13.61 24.64 46.43	211 55.24
	Commission	0	0	6 1.57 28.57 7.14	7 1.83 33.33 6.54	3 0.79 14.29 3.80	5 1.31 23.81 4.46	21 5.50
	Mayor Council	1	0	31 8.12 20.67 36.90	37 9.69 24.67 34.58	27 7.07 18.00 34.18	55 14.40 36.67 49.11	150 39.27
	Total			84 21.99	107 28.01	79 20.68	112 29.32	382 100.00

City Government Form by Number of Meetings Scheduled

Frequency Missing = 12

Working with ODS Graphics

Graphics are an indispensable part of statistical analysis. Graphics reveal patterns, identify differences, and provoke meaningful questions about your data. Graphics add clarity to an analytical presentation and stimulate deeper investigation.

SAS 9.2 introduced the Graph Template Language (GTL), a powerful new language for defining clear and effective statistical graphics. The GTL enables you to generate various types of plots, such as model fit plots, distribution plots, comparative plots, prediction plots, and more.

The GTL applies accepted principles of graphics design to produce plots that are clean and uncluttered. Colors, fonts, and relative sizes of graph elements are designed for optimal impact. By default, the GTL produces PNG files, which support true color (the full 24-bit RGB color model) and enable visual effects such as anti-aliasing and transparency. A PNG file retains a small file size. GTL statement options enable you to control the content and appearance of the graphic down to the smallest detail.

The GTL is designed to produce graphics with minimal syntax. The GTL uses a flexible, building-block approach to create a graph by combining statements in a StatGraph template. StatGraph templates are defined with the TEMPLATE procedure.

You can create custom graphs by defining your own StatGraph templates. To create a custom graph, you must perform the following steps:

1. Define a StatGraph template with the TEMPLATE procedure.
2. Use the GTL to specify the parameters of your graph.
3. Associate your data with the template using the SGRENDER procedure.

With just a few statements, you can create the graphs that you need to analyze your data. For example, you can create the following model fit plot with these statements:

```
proc template;
define statgraph mytemplate;
beginGraph;
   entrytitle "Model Weight by Height";
   layout overlay;
   bandplot x=height limitupper=upper limitlower=lower;
   scatterplot y=weight x=height;
   seriesplot y=predict x=height;
endlayout;
endGraph;
end;
run;

proc sgrender data=sashelp.classfit
              template=mytemplate;
run;
```

Display 7.5 *Model Fit Plot Using MyTemplate and Sashelp.Classfit*

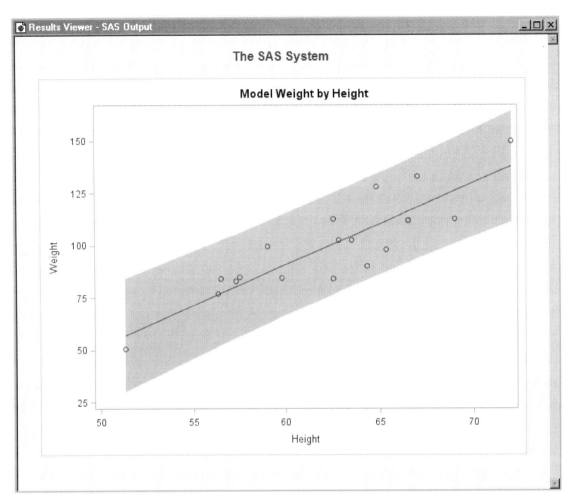

The previous example defines a StatGraph template named MyTemplate, which uses values from the data set Sashelp.Classfit. This data set contains the data variables HEIGHT and WEIGHT and precomputed values for the fitted model (**Y=PREDICT**) and confidence band (**limitupper=upper limitlower=lower**). The SGRENDER procedure uses the data in Sashelp.Classfit and the MyTemplate template to render the graph. (This example is member GTLMFIT1 in the SAS Sample Library.)

The following two graphics are more examples of what you can do with ODS graphics:

Figure 7.1 PROC SGSCATTER (SAS) with LISTING Style

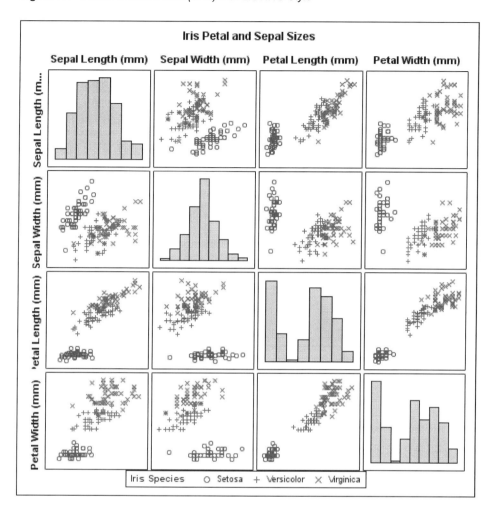

Figure 7.2 *Custom Template Rendered with PROC SGRENDER (SAS) and a Custom Style*

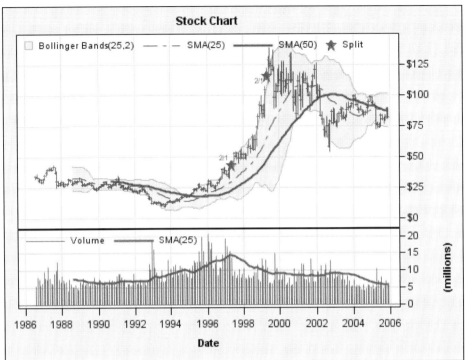

Advanced Features of the DOCUMENT Procedure

Overview

In Chapter 3, "Creating an ODS Document," on page 21, you learned how to manipulate output objects using the Documents window. You can use PROC DOCUMENT statements to accomplish all of the tasks that you completed using the Documents window and much more. For complete documentation on the DOCUMENT procedure, see Chapter 8, "The DOCUMENT Procedure," in *SAS Output Delivery System: User's Guide*.

The combination of the ODS DOCUMENT statement and the DOCUMENT procedure enables you to store a report's individual components. You can then modify and replay the report. The ODS DOCUMENT statement stores the actual ODS objects that are created when you run a report. You can use the DOCUMENT procedure to rearrange, duplicate, or remove output from the results of a procedure or a database query without invoking the procedure or database query from the original report. You can use the DOCUMENT procedure to do the following:

- Transform a report without rerunning an analysis or repeating a database query.
- Modify the structure of the output.
- Display output to any ODS output format.
- Navigate the current directory and list entries.
- Open and list ODS documents.
- Manage output.

With the DOCUMENT procedure, you are not limited to regenerating the same report. You can change the order in which objects are rendered, the table of contents, the templates that are used, macro variables, and ODS and system options.

Working with the DOCUMENT Procedure

To create an ODS document, you can use the Documents window or the "ODS DOCUMENT Statement" in *SAS Output Delivery System: User's Guide*. The following code creates the ODS document Work.Prddoc within a document store:

```
ods listing close;
proc sort data=sashelp.prdsale out=prdsale;
    by Country;
run;

ods document name=work.prddoc(write);

proc tabulate data=prdsale;
    by Country;
    var predict;
    class prodtype;
    table prodtype all,
    predict*(min mean max);
run;

ods select ExtremeObs;

proc univariate data=prdsale;
    by Country;
    var actual;
run;

ods document close;
```

The following display shows the ODS document Work.Prddoc and its contents. To view the Documents window, submit the following command on the command line:
odsdocuments

Display 7.6 *SAS Documents Window Showing Work.Prddoc Document and Documents Icon*

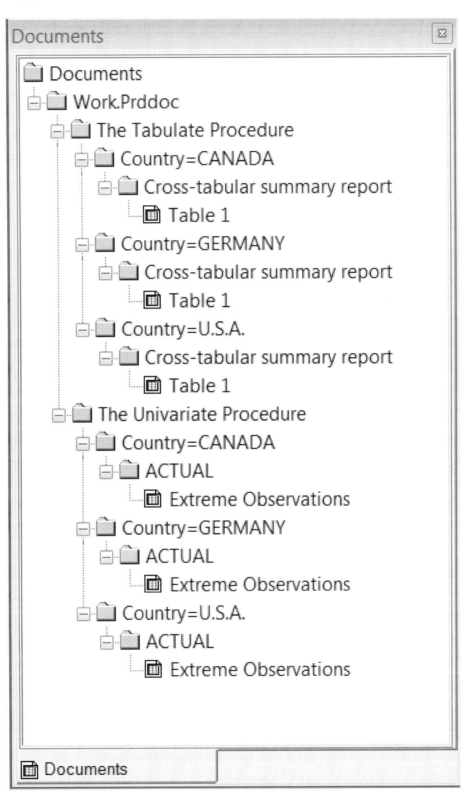

The following display shows the properties of Table 1. You can see the document name and the document path, as well as other information.

Display 7.7 Table Properties for Table 1

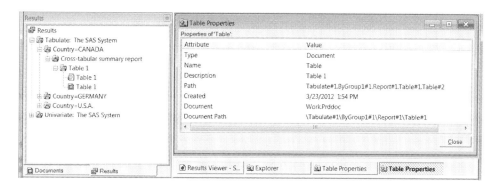

An ODS document store is not a SAS data set, as you can see by the Document icon in the previous display. The Work.Prddoc document was written to the Work library. If it had been written to a permanent location (such as `c:\temp\output`), in Windows Explorer, the document store would have a file extension of SAS7BITM.

After you have created a document with the ODS DOCUMENT statement, you can use the LIST statement in a PROC DOCUMENT step to view the contents of your document. The LIST statement enables you to look at the object list and folder structure in the ODS document. The following code creates a list of all levels of the Work.Prddoc document:

```
proc document name=work.prddoc;
    list / levels=all;
run;
quit;
```

The LIST statement can list what is in an entire document or in just one of the entries. For more information about the LIST statement, see Chapter 8, "The DOCUMENT Procedure," in *SAS Output Delivery System: User's Guide*.

In the following figure, every folder icon in the Results window corresponds to an item with a type of Dir in the LIST statement output. Every table created by a procedure corresponds to an item with a type of Table in the LIST statement output.

Figure 7.3 PROC DOCUMENT List Statement Output Compared to Results Window

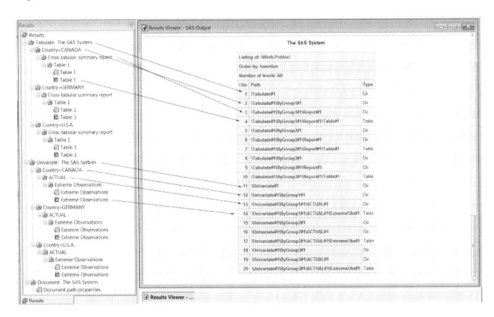

ODS and the DATA Step

If you are writing DATA step reports, you are already using ODS. HTML output, which is the DATA step output, is routed through ODS by default. For more than 20 years, SAS users have been able to create highly customized reports as simple LISTING output, which uses a monospace type font. With the advent of ODS, SAS users have a broad range of choices for printing customized DATA step reports.

- You can produce DATA step reports in many different formats, such as HTML, RTF, PS (PostScript), or PDF.

- You can create the report in multiple formats at the same time.

- You can produce the report in different formats at a later time without rerunning the DATA step.

To take advantage of these enhanced reporting capabilities, you can combine DATA step programming with the formatting capabilities of ODS. To create PDF output, for example, start with the DATA step tools that you are already familiar with.

- the DATA _NULL_ statement

- the FILE statement

- the PUT statement

Then, add a few simple ODS statements and options. You can choose from several ODS formatting statements to format the output in other presentation styles, such as HTML, RTF, and PS. For more information about ODS statements, see the "Introduction to ODS Language Statements" in Chapter 5 of *SAS Output Delivery System: User's Guide*.

Here are the basic steps for using ODS with the DATA step to produce reports with enhanced formatting:

Table 7.1 Steps to Produce Enhanced ODS Output with the DATA Step

Steps	Tools	Comments
Specify formatting for your output.	ODS formatting statements can specify formats such as listing, HTML, RTF, PS, and PDF.	You can produce output in multiple formats at the same time by specifying more than one format. *Note:* If you want only the default output, then you do not need a destination ODS statement.
Specify structure.	The ODS option in the FILE statement lists the variables and the order in which they appear in the output.	Additional suboptions give you even more control over the structure.
Connect the data to the template.	The FILE PRINT ODS statement creates an output object by binding a data component to a table definition (template).	You can specify other details by using ODS suboptions in the FILE PRINT ODS statement.
Output data.	The PUT statement writes variable values to the data component.	A simple way to output all variable values is to use PUT _ODS_.

Use ODS statements to specify how you want ODS to format your output (for example, as HTML, RTF, or PDF). Then, in the DATA step, use the FILE PRINT ODS and PUT statements with appropriate ODS-specific suboptions to produce your report.

The PUT statement writes variable values. The FILE PRINT ODS statement directs the output.[1] You can use ODS to produce the output in multiple formats and to produce output at a later time in different formats without rerunning the DATA step.

You control the formatting that is applied to your reports using ODS formatting statements. They open and close ODS destinations, which apply formatting to the output objects that you create with ODS and the DATA step.

Here is a list of topics and sources for additional information:

Table 7.2 Where to Find More Information about How to Use ODS in the DATA Step

Topic	Where to learn more
ODS formatting statements	Chapter 6, "Dictionary of ODS Language Statements," in *SAS Output Delivery System: User's Guide*

---

[1] If you do not specify a FILE statement, then the PUT statement writes to the SAS log, by default. If you use multiple PUT and FILE statements, then in addition to creating ODS-enhanced output, you can write to the SAS log, to the regular DATA step output buffer, or to another external file in the same DATA step.

Topic	Where to learn more
ODS destinations	"Understanding ODS Destinations" in Chapter 3 of *SAS Output Delivery System: User's Guide*
How ODS works	"Overview of How ODS Works" in Chapter 3 of *SAS Output Delivery System: User's Guide*

Where to Go from Here

Creating statistical graphics with ODS:
 For reference information about the GTL, see the *SAS Graph Template Language: Reference*.

Creating statistical graphics with PROC TEMPLATE and GTL:
 For usage information about PROC TEMPLATE and the GTL, see the *SAS Graph Template Language: User's Guide*.

Managing the various templates stored in template stores:
 For reference information about PROC TEMPLATE statements that help you manage and navigate ODS templates, see "TEMPLATE Procedure: Managing Template Stores" in Chapter 10 of *SAS Output Delivery System: User's Guide*.

Modifying an existing style or creating your own style:
 For reference information about style definition statements in PROC TEMPLATE, see "TEMPLATE Procedure: Creating a Style Template" in Chapter 13 of *SAS Output Delivery System: User's Guide*.

Creating and modifying ODS tabular output:
 For reference information about tabular template statements in PROC TEMPLATE, see Chapter 14, "TEMPLATE Procedure: Creating Tabular Output," in *SAS Output Delivery System: User's Guide*.

Modifying markup language tagsets that SAS provides or creating your own tagsets:
 For reference information about the markup language tagset statements in PROC TEMPLATE, see "TEMPLATE Procedure: Creating Markup Language Tagsets" in Chapter 15 of *SAS Output Delivery System: User's Guide*.

Using ODS with the DATA step:
 For information about using ODS with the DATA step, see Chapter 4, "Using ODS with the DATA Step," in *SAS Output Delivery System: User's Guide*.

Index

B
Base SAS reporting procedures 9
benefits of ODS 2
BODY= option
   ODS destination statements 39

C
components of ODS 2
CONTENTS= option
   ODS PDF statement 54
creating output
   combined program example 45
   Excel 45
   HTML 43
   PDF 42
   RTF 40
crosstabulation tables
   modifying 69
custom reports, example 17
customizing output 1, 49
   combined program example 59
   Excel 57
   HTML 56
   PDF 54
   RTF 51
   TABULATE procedure 52

D
data component 14
DATA step
   ODS and 78
   ODS reports with 79
default output, example 17
destinations
   See ODS destinations
DOCUMENT procedure
   advanced features 74
   capabilities 4
   compared to Documents window 38

documents
   See ODS documents
Documents window 5, 11
   compared to DOCUMENT procedure 38
   opening 21
   selecting output objects 29
   viewing output object labels 27

E
enhanced HTML
   creating output 43
Excel output
   creating 45
   customizing 57

F
features of ODS 1
FILE PRINT ODS statement 79
FILE= option
   ODS destination statements 39

G
global statements 5
Graph Template Language (GTL) 70
graphical output objects 2
graphics 70
GTL (Graph Template Language) 70

H
HTML output
   creating 43
   customizing 56

I
identifying output objects 25

integrating output with applications 39
item stores 14

N
NAME= option
  ODS DOCUMENT statement 21

O
ODS block 39
ODS destination statements 5
ODS destinations 14
  categories of 8
ODS DOCUMENT CLOSE statement 21
ODS DOCUMENT statement 4, 21, 74
ODS document stores 77
ODS documents 14
  creating, example 21, 75
  loading output 33
ODS EXCLUDE statement 29
ODS formatting statements 79
ODS global statements 5
ODS Graphics 70
  output 68
ODS HTML statement
  creating output 43
  customizing output 56
ODS PDF statement
  creating output 42
  customizing output 54
ODS RTF statement
  creating output 40
  customizing output 51
ODS SELECT statement 28
ODS TAGSETS.EXCELXP statement
  creating output 45
  customizing output 57
ODS templates 15, 65
  creating with TEMPLATE procedure 10
  modifying style definitions 66
  modifying table templates 66
ODS TRACE statement 25
ODS windows 11
output
  creating default, example 17
  creating Excel 45
  creating HTML 43
  creating PDF 42
  creating RTF 40
  customizing 1, 49
  customizing Excel 57
  customizing HTML 56
  customizing PDF 54
  customizing RTF 51

customizing TABULATE procedure output 52
  formatting for third-party software 5
  integrating with applications 39
  loading into ODS document 33
  location 4
  ODS Graphics 68
output control statements 5
output objects 2, 15
  identifying 25
  selecting 28
  selecting in Documents window 29
  templates 10
  viewing in Documents window 27

P
PATH= option
  ODS HTML statement 43
PDF output
  creating 42
  customizing 54
PDFTOC= option
  ODS PDF statement 54
PNG files 71
PRINT procedure 9
processing for ODS 3
PUT statement 79

R
REPORT procedure 9
RTF output
  creating 40
  customizing 51

S
SAS formatted destinations 8
SAS styles 49
selecting output objects 28
STARTPAGE= option
  ODS RTF statement 51
StatGraph templates 68, 71
style definitions
  modifying 66
STYLE= option 51
  ODS HTML statement 56
  ODS PDF statement 54
  ODS RTF statement 51
  ODS TAGSETS.EXCELXP statement 57
  TABULATE procedure 52
styles provided by SAS 49

T

table templates 15
  modifying 66
tabular output objects 2
TABULATE procedure 9
  customizing output 52
tagsets
  creating 68
  list of 3
Template Browser window 13
TEMPLATE procedure 65
  creating and modifying ODS templates 10
  creating tagsets 68
  modifying crosstabulation tables 69
  modifying style definitions 66
  modifying table templates 66

template stores 14
templates
  See ODS templates
Templates window 12
  opening 49
third-party formatted destinations 8

U

usage of ODS 4

W

windows for ODS 11
WRITE option
  ODS DOCUMENT statement 21

Made in the USA
Lexington, KY
05 February 2014